Mohammed RADOUANI

Digital Engineering

Mohammed RADOUANI

Digital Engineering

AERONAUTICAL MULTIPHYSICS SYSTEM

ScienciaScripts

Cover image: www.ingimage.com

This book is a translation from the original published under ISBN 978-620-6-72537-4.

Publisher:
Sciencia Scripts
is a trademark of
Dodo Books Indian Ocean Ltd. and OmniScriptum S.R.L publishing group

120 High Road, East Finchley, London, N2 9ED, United Kingdom
Str. Armeneasca 28/1, office 1, Chisinau MD-2012, Republic of Moldova, Europe
Printed at: see last page
ISBN: 978-620-8-21260-5

Contents

^Modelling and **Simulation** ^Multiphysics

Aircraft Brake Distributor

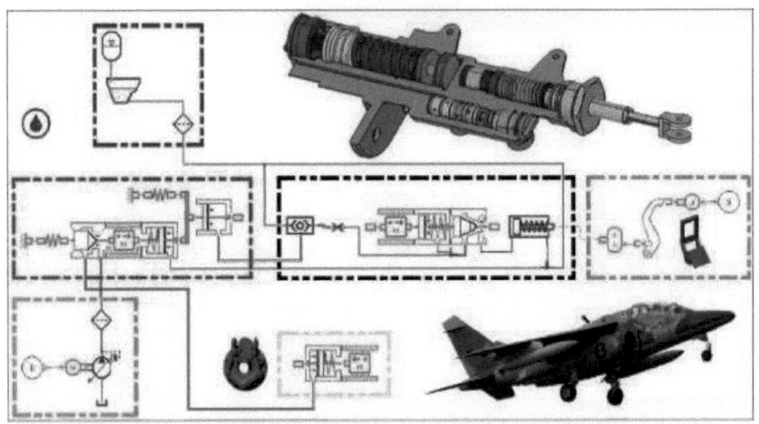

Resume

In aeronautics, aircraft braking systems are considered critical because of their significant influence on the dynamic behaviour and safety of the aircraft during taxiing. They are multiphysical in nature, using the hydraulic and electrical generations to develop mechanical braking friction. The aircraft braking system generally comprises two independent circuits: a normal hydroelectric circuit fitted with an anti-skid system controlled from the rudder pedals, and an emergency hydromechanical circuit.

The development of virtual prototypes of these systems is essential because they help both industry and scientists to validate experimental results and predict maintainability. It requires the simultaneous mastery of several disciplines: mechanics, thermics, hydraulics, tribology, etc.

In this context, the study focuses on the development of a cooperative physical geometric simulation (co-simulation) approach for multidisciplinary systems. The aim is to offer a range of advantages by combining the speed and robustness of the 1D physical model simulation with the sophistication and visualisation of the 3D-CAD geometric model.

The implementation concerns the study of the hydromechanical mechanism that controls, reproduces and regulates aircraft braking pressure. The aim is to develop a model that considers the main phenomena and parameters involved in the operation of the system.

Keywords: aeronautics; aircraft braking; multiphysics; CAD; moderation; cosimulation; AMESim.

CHAPTER 1

GENERAL INTRODUCTION

1- Context

In high-tech industries (aeronautics, maritime, rail, automotive, etc.), users are increasingly demanding high-performance products that simultaneously provide several functions in terms of efficiency, safety and assistance. This demand is prompting manufacturers to update their product range in order not only to meet the increasingly complex requirements of specifications, but also to counter the fierce competition on the global market. With the same objective in mind, designers have shown a strong tendency towards multi-disciplinary systems covering mechanical, hydraulic, electrical, thermal, pneumatic, IT and other fields. These systems are also referred to as "multiphysical" or "multidisciplinary" systems.

In aeronautics, the majority of aircraft systems are critical, with varying degrees of safety. Indeed, their malfunction can have harmful effects on people (crew, passengers or people on the ground), material assets (the aircraft, the airport, etc.) and the environment. These systems are classified according to their functions:

Engine systems: these are the accessories that generate electricity and fuel for the aircraft engine. In addition to the engine, they include the ignition system, the fuel supply system and the control system;

Avionics systems: these are electronic and electrical in nature, and include navigation, communication, radar and weapons systems;

Cabin safety systems: these are comfort and assistance systems for people on board. They are designed in pneumatic, mechanical, thermal and electrical disciplines. They include air conditioning, pressurisation, oxygen, demisting, defrosting and ejector seat systems;

Flight control systems: are of a hydromechanical or electro-hydromechanical nature and are used to operate the aircraft's control surfaces in order to control its trajectory and aerodynamic configuration around the aircraft's three axes (pitch, roll, yaw). They include the elevator, rudder, aileron and horizontal stabilizer control systems;

Utility systems: These are of a hydromechanical or electro- hydromechanical nature. They help control the dynamic evolution of the aircraft on the ground. They include the landing gear, aerobrakes, nosewheel steering and aircraft braking systems.

These systems must be highly reliable in terms of their ability to carry out their function during their service life, essentially while the aircraft is being flown.

In this context, the aim of assessing the performance of an aeronautical product at all stages of its life cycle, particularly during design and validation, is to achieve the quality and safety objectives expected of this product. The development of a multiphysics product must not only meet standards for efficiency, but also for increased reliability. As a result, monitoring its performance becomes an arduous necessity to satisfy, especially when added to the commercial desiderata of lower cost and shorter lead times, while contributing to the preservation of environmental resources.

Our work is part of the same approach to evaluating the efficiency and reliability of aircraft systems, and is part of the proposal for a federative project by the Scientific and Technical Research Unit of the Royal Air Force (FRA). The theme of this project is the virtual prototyping of aircraft systems, while developing approaches and methods dedicated to this objective. These prototypes will make a significant contribution to analysing and improving the maintainability of aircraft system components.

2- Problems

The development of numerical models of physical products remains an essential asset in the relevant analysis of system behaviour. These virtual models make an important economic and environmental contribution, as they avoid the need for expensive materials testing and reduce the time required for functional and integration tests.

Geometric modelling contributes to the development of 3D digital models of the products that make up systems. In this way, it generates a virtual conceptual 3D parametric representation that can be used to analyse, control and simulate certain kinematic and dynamic behaviours of mechanical parts separately. However, this modelling cannot generate the simulation of the evolution over time of the physical parameters of multidisciplinary systems comprising hydraulic, pneumatic or electrical components.

Physical modelling enables the evolution of these parameters to be simulated, providing an analysis of the magnitudes of systems and their components. Work in this field has led to the emergence of a large number of 1D numerical models, often specific to a particular field (mechanical, hydraulic, thermal, etc.). However, this modelling has the disadvantage of developing multiphysics models and the inability to visualise the dynamic and kinematic behaviour of the 3D assembly in the same simulation interface.

Admittedly, both modelling processes tend to reduce costs and design times by limiting testing through the virtual study of different solutions and by analysing system failures to contribute to the desired efficiency and safety. However, this evaluation is incomplete and separated from one process to another. As a result, there is a lack of simultaneous conceptual analysis that can identify the different physical parameters of the system simulation by visualising its behaviour and its kinematic and dynamic 3D animation. Therefore, in order to keep pace with the increasing demands of safety, efficiency and technical and economic constraints, it is necessary to consider unified 1D modelling methods for all trades. This will make it possible to design a new approach to concurrent geometric and physical simulation of multiphysical systems using appropriate software.

3- Contributions and objectives

The aim of this study is to develop a multiphysics physical modelling approach leading to 3D geometric - 1D physics co-simulation (cooperative simulation). This approach will make it possible to study, with animation
and visualisation, the dynamic and kinematic behaviour of systems and their multi-disciplinary components, taking account of variations in the various functional parameters.

The aim is to develop a virtual physical-geometric prototype for analysing and optimising the behaviour of these systems. In fact, we will present a physical modelling process dedicated to the development of 1D numerical models of multiphysical systems which will serve to interconnect with the 3D geometric model via an interface block enabling the co-simulation and analysis of the behaviour of these complex mechanisms by considering the parameters of the main phenomena involved in the operation of the system.

Our approach will be based on a study to identify the operating characteristics and the geometric and physical properties of each component of the multiphysics system. The use of this study will help us to develop the 1D physical model by adopting the component-oriented method and the 3D geometric model.

Analysis of the results of numerical simulations of the 1D model will enable us to validate and identify the behaviour of the various components and their interactions. Subsequently, a geometric-physical co-simulation will enable us to visualise the dynamic behaviour of the system components while analysing the various physical parameters associated with its operation. This analysis will be used to optimise the system's behaviour and predict the maintainability of the components and the entire system. This approach is applicable to multiphysical systems, in particular aircraft systems.

This multidisciplinary modelling and simulation approach will be applied to fighter aircraft braking systems. They are decisive in the control of the aircraft during taxiing, since they are essential during the various landing and taxiing stages right up to the aircraft's final stop.

4- Reading plan

The structure of our study is based on four main chapters.

The second chapter, which is fundamentally bibliographical, presents a short introduction to the multidisciplinary nature of systems and the classical design methodology for multiphysical systems. This is followed by a review of the main approaches to modelling a multiphysics system and the various associated tools. This literature review highlights the problem of modelling and simulating a multiphysics system.

The third chapter will describe the hydraulic and electrical generation systems, as well as the standard hydraulic components that make up an aeronautical multiphysics system. In addition, to make our approach more concrete, we will carry out a detailed study of two aircraft braking systems, namely: the first relates to the hydromechanical system and the second concerns the electro-hydromechanical system equipped with an anti-skid system. This study covers the operating principle and the geometric and functional properties of each of the components of these systems. As a result, we are developing 3D geometric models of the multiphysics aircraft braking subsystems.

The fourth chapter outlines the approaches to 1D physical modelling of a multiphysics system. It then presents the modelling methodology that will be adopted, in this case the component-oriented approach. Using the functional study carried out on the systems under study, we will develop 1D physical models of these systems. The proposed modelling is consistent with all trades and takes into account all aspects related to the analysis of the parameters of a multiphysical system.

The fifth chapter will be devoted to parameterising the simulation of the 1D physical models with a view to validating them. This validation involves simulating the functional test described by the aircraft manufacturer and evaluating the functional parameters in relation to those required by this test. Following a brief presentation on the concepts of co-simulation and the numerical model, we will focus on the 'interfacing' between the modelling tools for multiphysics systems. To this end, we will develop a co-simulation method between the 1D physical model and the 3D geometric model of a multiphysics braking mechanism for the simultaneous visualisation and analysis of the dynamic and kinematic behaviour of the system. Finally, we will look at the system response time parameter in order to optimise this parameter and contribute to the system's efficiency.

The conclusion of this study will focus on the main results obtained and the achievements developed throughout the various chapters. In addition, a number of research avenues that could be explored to complement and extend this work will be presented.

CHAPTER 2

MULTIPHYSICS SYSTEMS - DESIGN & MODELLING -

1- Introduction

The globalisation of markets and continuous economic fluctuation are forcing industries to be more competitive by constantly renewing their product range to distinguish themselves from competitors. The vehicle industry, and aircraft in particular, has seen technological advances in the design of component systems to keep pace with these developments and ensure that they function properly. This has led to the emergence of more efficient products that encompass several disciplines at once (hydromechanics, electrothermics, electro-hydromechanics, etc.) to meet the new standards for safety and efficiency. The systems that make up these products are increasingly complex and are known as 'multiphysical systems' or 'multidisciplinary systems'.

This chapter presents a state of the art on the multidisciplinary nature of systems and a revolution in the methodology for designing and modelling multiphysical systems. It is divided into two parts:

In the first section, we present the concept of modelling, the main approaches to physical modelling of a multiphysics system and the various associated tools. In the second part, in the light of this literature review, we identify the advantages and also the limitations of each approach in order to highlight the problem of modelling and simulating a multiphysics system.

2- General points: systemics and multidisciplinarity

2.1- Systemic approach *

Before delving deeper into the problem of this research work, it is necessary to clarify a certain number of ideas and concepts related to the notion of system and system engineering.

2.1.1- System concept

PENALVA's usual definition of a system is a set of organised means directed towards a single end [1]. To be more precise, BREAS defines a system as an organised global unit of interacting elements, functioning and evolving according to a purpose, immersed in an environment that acts on it and on which it acts [2]. This definition is based on the archetype proposed by LE MOIGNE: the General System, which is described by an Action (an interweaving of

Systemics is defined as a new discipline which brings together the theoretical, practical and methodological approaches relating to the study of what is recognised as being too complex to be approached in a reductionist manner, and which poses problems of boundaries, internal and external relationships, structure, laws or emergent properties characterising the system as such, or problems of the mode of observation, representation, modelling or simulation of a complex totality. These are problems of observation, representation, modelling or simulation of a complex totality. These are problems of a system (or system of actions) in an environment ("lined" with processes) for a few projects (purpose, teleology) functioning (doing) and transforming (becoming) [3].

A system can also take the form of a structured set of abstract elements, i.e. a theoretical construction formed by the intellect on a given subject. In this way, the system is qualified by the observer, who attributes properties to it. According to PENALVA, any situation that is difficult to understand, anticipate or control is considered complex [1]. Complexity is not an intrinsic characteristic of a system. Complexity is attributed to the system by an observer. Depending on the observer's intentions, the system may be [4] :

An autonomous system in an environment ;

A subsystem as an integrated part of a system;

A supra-system dominating other systems without encompassing them;

A meta-system encompassing systems of all kinds and giving them meaning.

2.1.2- *Systems engineering*

In the face of numerous challenges (competition, globalisation, pressure on costs, deadlines, quality) and increasing constraints (financial, political, environmental), it is necessary to master the design of complex systems and products. Systems engineering was developed in response to this need to understand and master complex systems.

According to IEEE, "Systems engineering is a cooperative and interdisciplinary approach to the progressive development and verification of a solution for the system, balanced over its entire life cycle, satisfying the expectations of a customer and acceptable to all" [5]. According to MEINADIER, systems engineering is a collaborative, interdisciplinary problem-solving process that draws on knowledge, methods and techniques from science and experience to define a system that satisfies an identified need [6]. We can also define systems engineering as an approach based on heterogeneous conceptual and technical foundations that can be used to solve problems [7]. It is used in particular for :

To design, evolve and verify a system that provides a solution to an identified operational need in accordance with measurable efficiency criteria [8] ;

Satisfy the expectations (quality, innovation, performance) and constraints (deadlines, costs, performance, operational safety) of all its stakeholders and be acceptable to the environment;

Balancing and optimising all aspects of the overall economics of the solution throughout the system's lifecycle.

2.2- Multi-disciplinary systems

2.2.1- Multiphysics system

A multiphysics system is a system that integrates a number of physical disciplines: mechanics, hydraulics, computing, electricity, etc. (Figure 1). However, this integration of disciplines at the same time considerably increases the complexity of the system, even at the level of its design, modelling and study. This multi-disciplinarity favours efficiency in operation and also less clutter in terms of the multitude of functions they provide for the systems. Hence the emergence of design methods that enable designers to integrate all the disciplines simultaneously. Most industries are experiencing this evolution in their systems, particularly the automotive, rail and aerospace industries.

In this study, we assume that the types of complexity can occur together or separately. We will therefore incorporate them implicitly into our approach. Among the names assigned to the classification of systems which are also considered multiphysical, we cite: complex system and mechatronic system.

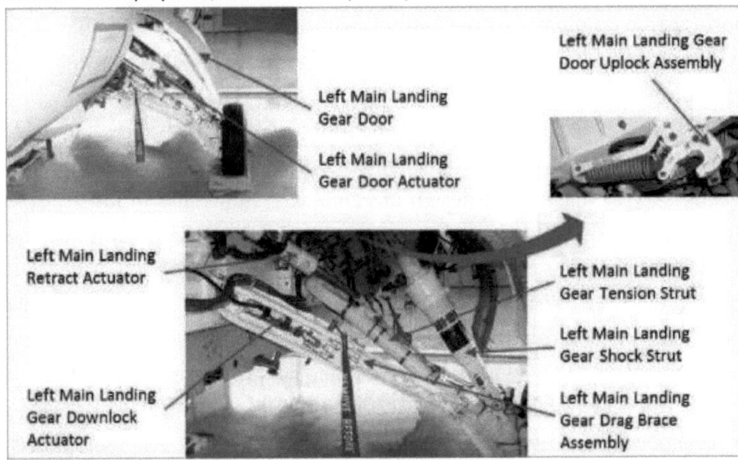

Figure 1: *Multiphysics system for fighter aircraft landing gear*

2.2.2- Complex system

By complex, we mean systems at the organic, structural and evolutionary level, comprising entities or agents interacting in a non-linear structural manner and carrying out activities with a view to achieving a given objective. This complexity is of several types [9] :

- **Structural:** the structure, organisation, number of entities and nature of the links are all taken into account together, which makes it difficult to model and process the associated information;
- **Algorithms:** the complexity here is intrinsic and is aimed at problems that are difficult to grasp and understand. In this case, it is necessary to use particular

methods of solution (for example: Heuristics);

- **Behavioural:** from this point of view, the evolution of the system, although using representation models and simple algorithms,

is unpredictable and results in the spontaneous emergence of complex forms and behaviours (e.g. deterministic chaos).

A complex system is, by its very nature, multi-technological (Figure 2), which necessarily means that its design is multi-disciplinary, from the analysis of the initial need right through to the overall digital mock-up defining the product to be industrialised 10]. It responds to a need defined by functional specifications:

throughout its life cycle;

by interacting and communicating with its environment;

based on a mechanical structure whose behaviour is managed algorithmically, using electronic and energy resources, ...

Figure 2. *Complex aircraft navigation systems*

2.2.3- *Mechatronic system*

Mechatronics, this neologism, is a multi-domain combination encompassing an approach aimed at the synergistic integration of different disciplines in the design and manufacture of a product with a view to increasing or optimising its functionality. This raises the question of what exactly 'mechatronics' is and which disciplines are involved.

Current definitions describe mechatronics as an interaction between electronics, mechanics and information technology [11]. BOLTON's 1999 definition [12] :

"A mechatronic system is not just a marriage of electrical and mechanical systems and is more than just a control system; it is a complete integration of all of them.

The definition used by the international journal IEEE, Transactions on Mechatronics [13, 14] :

"Mechatronics is the synergistic combination of mechanical engineering with electronics and intelligent computer control in the design and manufacturing of industrial products and processes.

A typical mechatronic system consists of an embedded system that controls a physical system through actuators, sensors, controllers, hardware and software computer interfaces (Figure 3). Various devices for sensing, acquiring and transferring information are involved between these components.

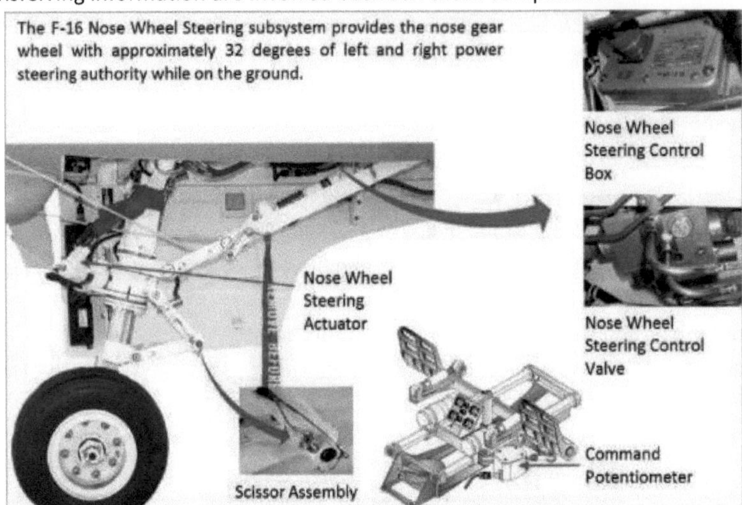

Figure 3: *Mechatronic steering system for F16 aircraft on the ground*

Historically, complex systems have become mechatronic. Here are a few milestones that will help you put mechatronics in context:

Before 1950, complex systems were essentially electromechanical assemblies. The 1950s saw the advent of semiconductors, and control and power electronics were born. In the 1960s and 70s, the emergence of reliable computers enabled the design of programmable control systems that were more powerful and flexible. In 1969, YASKAWA ELECTRIC CORPORATION (JAPAN) coined the term "mechatronic" to describe a system that closely integrates mechanisms and programmable electronics. More recently, the development of high-integration microcontrollers and Flash memories, as well as the use of advanced programming languages, have made it possible to design high-performance mechatronic systems in very short timescales. According to the NF E 01-010 standard published in 2008 [15], mechatronics is defined as :

"Mechatronics is an approach based on the synergistic integration of mechanics, electronics, automation and IT in the design and manufacture of a product with a view to increasing and/or optimising its functionality.

Today, mechatronic systems mainly draw on four disciplinary fields [16] (Figure 4):

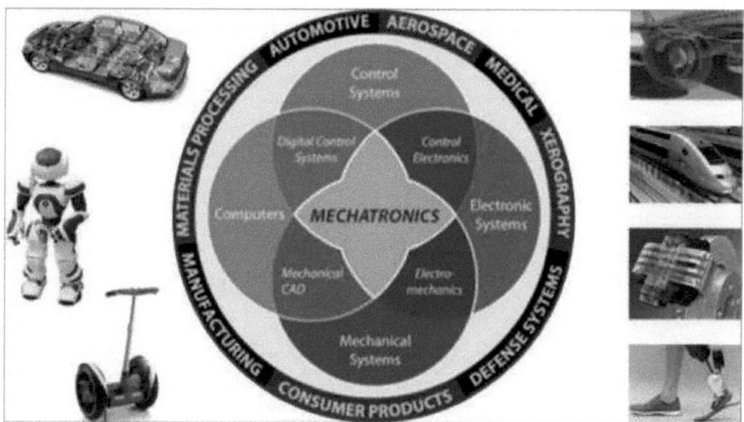

Figure 4. *Disciplines of a mechatronic system*

Mechanics: in the form of mechanisms and materials (static, dynamic, RDM), but also hydraulics and pneumatics.

Electronics: control and power (microcontrollers, power bridges, converters).

IT: (real-time, networks, signal processing).

Automatic control: (linear and non-linear servo-controls, control-command strategies).

3- Design of a multiphysics system

Product design consists of designing, developing and marketing a new product that differs from existing products and meets the needs of the user. This product or industrial system fulfils a function that is directly linked to the needs of a user customer, whereas the same function can be achieved using a multitude of systems.

The design objectives of a system are expressed in terms of the desired performance specifications. By definition, a perfectly successful design is one that accurately meets the design objectives (design criteria and specifications).

The need to satisfy the needs of the end-user customer calls for detailed thought and essential formalisation stages, particularly during the product design phase. Generally speaking, there are three levels of action involved in developing a product: the design office, which is responsible for product design, the methods office and the supervisory staff, who are responsible for implementing the design, and the workshop, which is responsible for manufacturing.

Design (also known as the study phase or the definition phase) is one of the phases in the life cycle of an industrial product (Figure 5).

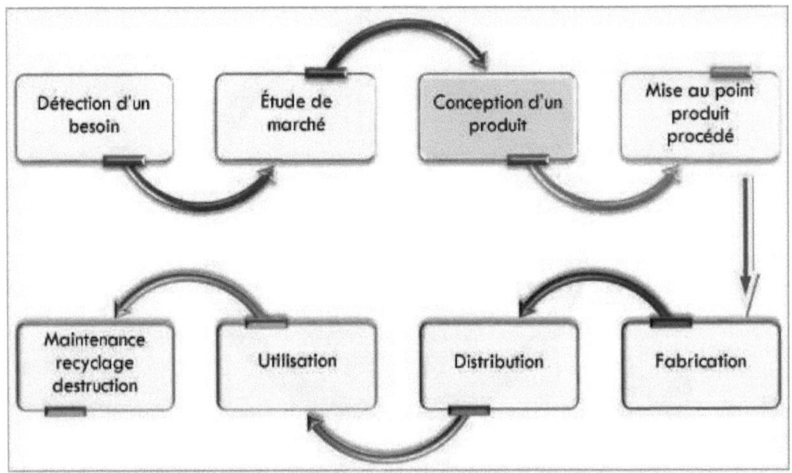

Figure 5. *Phases de cycle de vie d'un produit industriel*

3.1- Traditional design methodology

The classic design of complex systems usually involves two decompositions:

Sequential: sequential decomposition of design according to the product life cycle. These are studied one after the other:

product study/design;

the manufacturing process ;

product distribution (commercial aspect);

product maintenance.

Professions: breaking down the product into "profession" sub-systems. Each trade intervenes sequentially according to a procedural approach to design:

mechanisms separately by discipline ;

motorisation ;

control and command ;

of energy.

The 'sequential and job-oriented' approach was adopted for the design of multidisciplinary systems such as electromechanical systems. It consisted in breaking down the multi-physical system into several sub-systems (each depending on a single physical discipline) and studying these sub-systems separately. In effect, the mechanic designs the mechanical mechanisms and the electronics engineer draws up the functional specifications for the machine (actuators, sensors, electronic cards, etc.) and the same goes for the computer scientist, who proceeds with the conceptual implementation of the algorithms in a predefined automatic.

This design method does not allow us to focus on optimising the multi-technology system and its overall integration. This requirement can no longer be met by breaking things down by discipline. Indeed, from the properties of the different parts, it is not trivial to deduce the behaviour of the whole and thus ensure its maximum efficiency [6]. On the other hand, in the sequential approach, even if each of the specialists works as efficiently as possible, their isolation means that they only have a partial idea of the design of the overall system, which can lead to interfacing problems between the subsystems. This approach is no longer compatible with the demands of today's markets: higher quality, lower costs, shorter design and manufacturing times, miniaturisation and integration, more appropriate maintenance, recycling, globalisation of the markets, and so on. There is therefore a need for a new approach to the design of complex systems, enabling the product to be optimised as a whole rather than just in part.

A combination of different technologies can only be optimal if the following two conditions are met:

real communication between the various specialists has been put in place;

the interconnection of different disciplines is taken into account in the design process.

Hence the advent of methods that are rooted in the need for engineers and technicians from different fields to work much more closely together than ever before. It is now a question of knowing how to integrate the different technologies so that they interact with each other in the best possible way.

3.2- Multiphysics design approach

Multiphysics system design should consider the system as a whole in a concurrent manner rather than designing system aspects separately and sequentially. Such a conception resorts to a new look at the design process itself, and to a formal consideration of the transfer of information and energy between the components of the system. It is in this context, in particular, that the notion of concurrent engineering has emerged, as used in PLM (Product Life Cycle Management). Also known as simultaneous engineering, the mechatronics approach or the systems approach [11].

3.2.1- Collaborative and concurrent engineering

Its aim is to place product design within its overall life cycle and it is the very essence of the mechatronics approach. It enables coupled functions to be designed by integrating mechanical, electronic and IT aspects [17]. This multi-disciplinary approach means that design must be based on the concept of co-development (or co-design) of the mechanical system, hardware, software and work interfaces, with the pooling of a technical database and an overall digital model of the system. It is the only resource that can be used by all the 'trade' tools, whether mechanical (finite elements, etc.), electrical (electromagnetic compatibility, etc.) or IT.

The basic concept of this new approach is communication between the various parties involved in the design of a multiphysics system. Teams made up of engineers with multi-disciplinary backgrounds work alongside more specialised trade teams, ensuring ongoing dialogue and guaranteeing the coherence of the various studies. This will make it possible to take into account, from the design stage onwards, the various phases in the system's life cycle (manufacture, control, maintenance, etc.), and to envisage innovative solutions that will result in a more precise, higher-performance product compared with a non-multiphysical product [18]. This justifies the application of this design synergy in many industries.

3.2.2- V design cycle

The V-cycle approach has been adopted as a benchmark for the design of multiphysics systems [19].

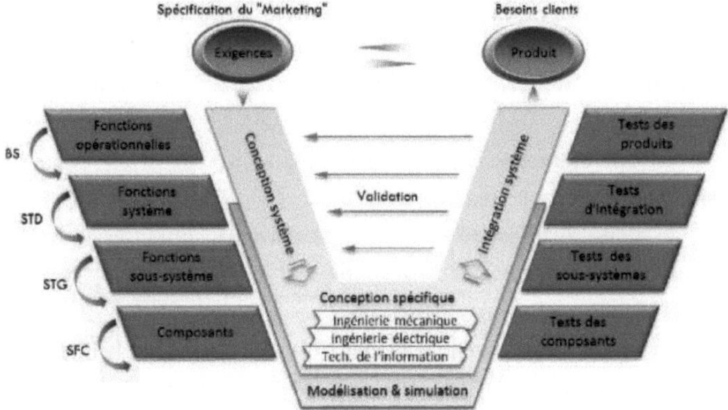

Figure 6. *V cycle for the design of a multiphysics system*

Or : BS (Besoin et Specification) ; STD (Specification Techniques Detaillees) ; sTG (Global Technical Specification); sFC (Component Manufacturing Specification).

Figure 6 illustrates the design process based on the principle of this cycle, which passes through the following three main stages:

System design phase ;

Specific design phase ;

System integration phase.

The starting point of the design cycle is the requirements specification, which expresses the expectations and needs to be met by the system to be developed.

The aim of the system design phase is to define, from the requirements of the specifications, a solution concept which describes the main logical and physical aspects of the future system. To do this, the system's main function is broken down into sub-functions. Each sub-function is assigned an operating principle or solution elements. The performance of each sub-function is checked in the context of the system.

The specific design phase is devoted to the more detailed definition of each element of the system solution. This stage is generally carried out separately by technical discipline (e.g. mechanical, electrical) and by the corresponding business specialists.

The system integration phase consists of assembling the elements or components produced separately in the previous phase, to gradually form the complete system. It is then possible to study the interactions between the various elements of the system, as well as the behaviour of the overall system. At the end of the cycle, the developed system is evaluated by comparing its characteristics and performance with the requirements of the specifications.

A complex system is rarely completed in a single cycle. In reality, a series of cycles are required. For example, in a first cycle, the system is functionally specified, the main concepts and operating elements are defined, roughly dimensioned and their characteristics and properties are verified in the context of the overall system. The result is generally a laboratory prototype. In a second cycle, the definition of the system is refined (fine sizing of components, detailed simulation of behaviour, etc.), culminating in a functional prototype. Depending on the progress of development and the complexity of the problems involved

In the event of such problems, a number of additional cycles may be required to reach the industrial production stage.

The project is thus broken down into several sub-projects at different levels. The teams in charge of designing the product will develop and test all possible solutions and the V cycle will be successively run through several times, totally or partially, until an optimum solution is found. For this reason, it is preferable to implement iteration loops as early as possible in the design cycle to accelerate it.

In short, multiphysics product design should not only involve designers, but also the people responsible for developing the process, manufacturing the product, marketing it, maintaining it and recycling it. In this respect, simultaneous rather than sequential design, taking into account the interactions between disciplines, is beneficial. Hence the importance of developing a multiphysics approach to modelling multidisciplinary systems.

4- Modelling a multiphysics system

4.1- Concept of a model

4.1.1- Definitions

Etymologically, the word "modele" comes from the Latin "modullus", a diminutive of "modus" (measure). Initially, this term was used in the field of architecture to designate an arbitrary measurement used to establish the ratios between the different parts of a building under construction. It was then used to designate any modelling activity: a model can be defined as an abstract description of a physical reality, which is intended to be usable in the design process.

The very general nature of this definition gives us a foretaste of "the world of modelling". Modelling is all the mathematical, numerical and computer tools that combine to build a virtual model (or computer model) of a real object. This object may be more or less complex, more or less schematised. It can be the fruit of the imagination, of a trend or rather a more or less exact solution to a given physical problem, or even a compromise between the two. By definition, a model is a simplification of reality, because if it had all the attributes of reality, it would no longer be a model but reality. Thus, the model will be an 'image' (and not a copy) of the original that meets the functional and performance characteristics previously defined by the observer or designer. The system under study is called the primary system. A model is a representation of reality and is called a secondary system.

The words 'model' and 'representation' are highly polysemic and are sometimes synonymous. The term "model" can be taken in an "ante" sense: an object to be imitated, a matrix; or in a "post" sense: an object reproducing a portion of reality (a representation). In either case, the model is a tool used to make or learn something [20].

4.1.2- *Types of models*

The stages in a computer modelling process produce different types of model. We will use the terms used by VANBERGUE [21]:

A domain model defines the data, behaviours and assumptions used in the target system. It is often described in natural language;

A design model is a formalised version of the domain model, in which the constraints created by the modelling tool are integrated. A design model is a communicable "paper" version of the operational model;

An operational model is the computer model itself, implemented in the machine and used as the basis for simulations.

Disciplinary models provide partial knowledge of a system observed from a certain point of view and on a certain scale. They provide a precise approach to a well-defined aspect of a problem, but only an approach that integrates different points of view and different scales can shed light on the functioning of the system as a whole. Hence the need for interdisciplinary models that relate to the modelling of systems generated by the management of different disciplines.

4.1.3- *Simulation models*

Simulation models - henceforth referred to simply as models - are simplified representations of systems evolving over time. They can be used for predictive purposes: starting from the current state of the system, and making assumptions about its evolutionary dynamics, a simulation can be used to describe the future states of the system for different scenarios. The relevance of predictive models is that they provide additional information once the parameters of a system have been established.

But in the field of multidisciplinarity, the systems considered are unpredictable by their complex nature. The dynamics of such systems emerge from interactions that are localised, interdependent and spread over different scales. They have strong characteristics of non-linearity and dependence on the history of the system.

Explanatory models, the aim of which is not to predict the future state of a system, but to shed light on the way it functions today. DORAN talks of the 'soft benefits' of modelling: highlighting a gap in knowledge, making a pedagogical contribution, a means of communicating and focusing thought [22].

4.2- System modelling

4.2.1- Modëlisation with structural dëcomposition

A model of a system is generally an abstract description of the system using a given theory or symbolic language. It is a simplified (model) or abstract (diagram, graphic representation, mathematical equations, etc.) representation of a physical system. Models can be used for a variety of tasks: in conceptual design, models representing alternative solutions help us to test the behaviour of a technological device before it is built; in a monitoring problem, a model can reveal a fault in the system.

Traditional modelling is based on a fixed, structural decomposition. Thus, the system (Figure 7) comprises:

an operating part (OP) grouping together all the mechanisms and actuators of the overall system;

a control unit (PC) which manages the overall behaviour of the system;

interfaces between the PO and the PC, as well as with the system environment (pre-actuators, sensors, Human Machine Interface, etc.).

Each of these parts is then modelled independently of the others, using incompatible tools (CATIA, MATLAB, AMESIM, SPICE, etc.), whether from a structural or behavioural point of view.

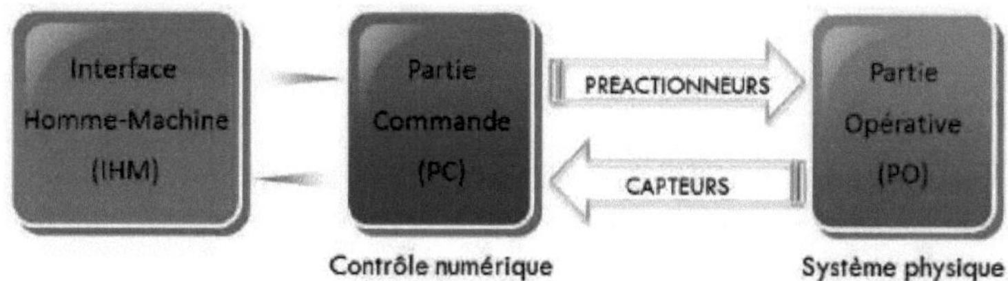

Figure 7. *PO-PC architecture*

This division separates the mechanics and control-command professions (electronics, mechanics, electromechanics and IT). As a result, it does not allow reliable modelling of the real multiphysics system, nor simple verification of the coherence of its behaviour.

4.2.2- System model abstraction levels

In most design cases, it is not possible to answer the questions that arise with a single model. Different models are needed to describe the different stages of design. Consequently, computer-aided design tools are adapted to model and calculate the corresponding models. These tools can be classified according to their role during the design process described by the V-cycle. Thus, the different design stages are associated with simulation tools that meet the different levels of abstraction required. This level of detail is commonly referred to by designers as the abstraction level [23].

Modelling can begin once the level of abstraction has been set by the designer and validated by all project collaborators, in accordance with the specifications. We have identified four levels of abstraction (Figure 8):

Functional level: this defines the functionalities that the product will have to offer (translation of the requirements expressed by the customer into Technical Specifications of Needs) and the life situations it will be confronted with. It involves naming the functions of the system under study and describing its successive states when an event is put into action. i.e. :

- conducting an "internal" functional analysis of a solution, from a structural and behavioural point of view;
- the characterisation of the solution by all the functional and non-functional specifications, in the form of constraints and relationships on the characteristic parameters of the system.

System level: This represents a physical view of the product in the sense that the architecture of the system (number of sub-systems needed to provide the various functions, nature of the interactions between them, etc.) is already outlined. Certain technological choices are therefore implicitly decided when the Global Technical Specifications are written.

At this level, the model describes the behaviour of the physical system and the controls. This representation includes "block diagrams" containing the system's behavioural parameters in the form of gains, tables, curves, delays, mathematical operators, etc. This type of description is called *signal flow* and is purely mathematical in nature.

Sub-system level: Each sub-system defined at the "system" level is seen as an organ of the initial product. In order to specify its design, each component must in turn be broken down into sub-systems (components).

New architectural and technological choices are documented this time in Detailed Technical Specifications. This is therefore a macroscopic level of modelling for which the components are interconnected by rules based on the exchange of power at the interfaces. These interfaces are called "ports", and the components are represented as "multiports".

Modelling at this level is based on the structure of the models in elements and their interfacing, and is distinguished from the 'system' level by the conservation of energy property. The latter is well known in different areas of physics: KIRCHHOFF's law in electricity, NEWTON's third law in mechanics, etc. The structuring approach of the multi-port language used contributes to exploiting all the properties of the models and to taking into account the technological aspects of the trades, which often condition the modelling assumptions.

Geometrical level: At this level, the models include both the parameters describing the 2D or 3D geometry, and the behaviour laws of the materials. The problem is defined by partial differential equations (PDEs) which are processed using a discretisation or mesh of the geometry (finite elements, finite differences, finite volumes, etc.).

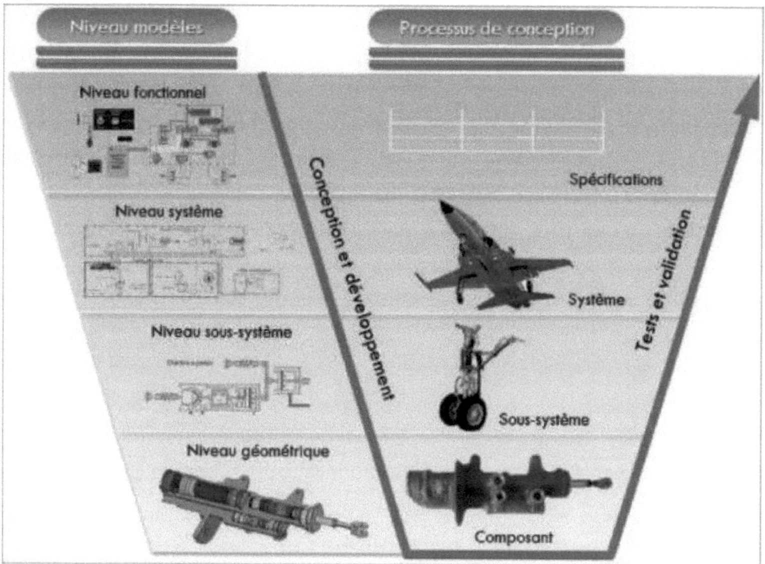

Figure 8. *Levels of abstraction associated with the V cycle*

The criteria we consider for this modelling depend on the type of verification we wish to carry out subsequently. Model expression languages have been defined specifically for the different criteria to be observed. We will list a few of these to give an idea of the diversity of the world of systems modelling:

- 1D physical modelling: representing the architecture of a system using modelling languages in the form of a set of components
interacting. The system will be described essentially from the point of view of its

structure: the intra-component characteristics as well as the inter-component characteristics (the interactions).

/ 3D geometric modelling: representing a system in three-dimensional space

4.2.3- *Virtual prototyping*

The levels of abstraction outlined above are associated with different types of modelling and simulation tools that define a "platform" for "virtual prototyping". In a virtual prototyping design problem, one of the most difficult tasks is the modelling stage, 24 , which requires :

find a level of abstraction appropriate to the design problem under consideration ;

be able to describe the various phenomena involved in the system and identify whether they fall within the required level of abstraction;

represent these different aspects in a unified way, even if they belong to different areas of physics (mechanics, electricity, hydraulics, etc.).

Clearly, the single model representing the virtual prototype for the entire platform is unrealistic. In fact, we can consider that there are almost as many models as there are questions, and tools that are better suited than others (or completely unsuited) to answering certain questions. This leads to an "integrated design environment" made up of different tools that must not be considered independently of each other in order to ensure the continuity of the design cycle.

The fundamental specifications for such an environment are based on the work carried out by KASPER, which highlights the following elements [25]:

Openness: In an industrial context, simulation techniques must not only meet the calculation needs of a single company, but must also enable models to be exchanged between equipment manufacturers and assemblers. This requires the use of simulation software. We can also note the standardisation efforts being made, such as [15], which represents a language for digital and analogue electronics.

Interdisciplinarity: The design of mecatronic or multiphysics systems requires an environment capable of supporting the modelling of components from different fields of physics and their integration with other models in order to describe the complete system.

Reusability: The use of well-structured and documented model libraries reduces time and effort for designers. This aspect is not limited to design in the sense that valid models can be used by manufacturing engineers who are interested in optimising manufacturing tolerances in order to reduce production costs while satisfying functional product specifications.

Portability: The design of complex products is often split between teams working on several sites. For this reason, the simulation environment must be accessible via Intranet or Internet and run on different computer platforms such as Windows and UNIX.

Adaptability: Many physical phenomena do not have a single mathematical representation. The fact that several levels of modelling are available, associated with clearly expressed hypotheses, and that the complexity can be easily chosen, ensures that the models can be adapted to the questions posed.

4.3- The contribution of moderation to design

The contributions of modelling are quite different depending on: the use of the model: research (model of a system intended for researchers), diagnosis (model of a system intended for those commissioning a project on the system), intervention (action research model); the objective of the model: understanding the processes of a system, gathering knowledge, exploring scenarios, etc; the realism of the model: simplified reality or abstraction; the content of the model: derived from industrial representations or scientific knowledge; and the modelling process itself: model produced by researchers or co-constructed with industry.

Despite the specific features of the models developed, modelling makes an active contribution to reducing the cost and time involved in design, understanding how the system works, and validating and optimising the model under study.

Cost and time reduction: The technical solutions in the specifications during product design must adequately meet the criterion of customer satisfaction. To verify this response, we measure the difference between the actual functions of the designed product and the expected specifications. On the one hand, this modelling reduces time and cost while improving product quality compared with the development of real mock-ups. On the other hand, it provides the opportunity to simulate, analyse and optimise the solution system in order to reduce the gap between the solution and what is expected.

Understanding how it works: Developing an abstract model that takes in the basic functionality of a physical component is very useful for understanding how it works. The whole art of modelling is therefore to ask the right questions a priori so that we have an initial executable model and we can master the behaviour of the component throughout the design process without changing the test vector.

Model validation: The main purpose of a model is generally to validate the characteristics of a certain part of the system or of the whole, in terms of functionality or performance. It is the definition of these characteristics and functionalities that helps to verify that a system exhibits an expected behaviour by observing its behaviour in the face of different stimuli (system tests). Hence the role of developing a product simulation model as a solution.

This model will not be valid outside the scope previously defined. In fact, by following a "Top Down" modelling method (from the most abstract level to the physical level), omitted details can be implemented in order to validate other characteristics of the system.

Model optimisation: The aim of a modelling process is to obtain a final model that meets simulation and verification requirements. The development of a system therefore results from the implementation of numerous interconnected modelling stages. These stages, based on the requirements of the specifications, converge towards virtual and then real prototyping of the system itself. One of the main advantages of modelling is that it is possible to specify an optimisation criterion to rank the possible solutions. This criterion will make it possible to choose from among all the possible solutions those that appear to be the most interesting. The criterion consists of maximising or minimising the value of an indicator constructed from the variables in order to reduce the set of acceptable solutions (Figure 9).

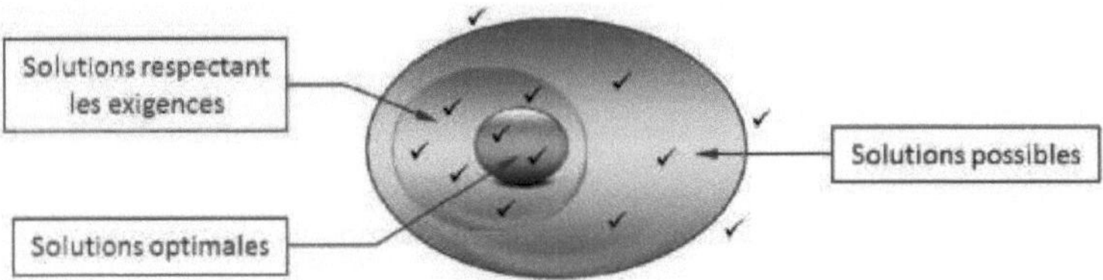

Figure 9. *Set of optimal solutions*

In short, it is clear that the development of a product model during the design phase plays a key role. However, it is not without its difficulties, since designers must define the point of view from which they wish to approach the product, the validation domain and the precision. Hence, it is essential to carry out integration work at the earliest stages of the design cycle. This requires a methodological approach involving modelling and simulation at the earliest stages of the design process.

5- Modelling approaches and tools

5.1- 3D geometric modelling tools

Geometric modelling contributes to the development of 3D digital models of the products that make up systems and assemblies. It generates a virtual conceptual 3D parametric representation that is used to analyse, control and simulate certain kinematic and dynamic behaviours of mechanical parts separately. It also enables the 3D geometric model to be refined and visualised (Figure 10).

Figure 10: *Examples of 3D geometric modelling of products*

It conditions the data structure and interactivity. Conversely, the type of final output can condition the model. Each model is characterised by :
data structure: description language, acquisition, interactive graphics system, cost (graphics pipeline),
the final appearance: degree of realism, speed of generation, degree of interaction: editing, modifications, calculations, analysis, deformations.
Dassault Systemes, one of the world's leading suppliers of 3D and Product LifeCycle Management (PLM) solutions, has placed the modelling of embedded systems at the heart of CATIA. It has chosen the MODELICA open standard as the basis for its open strategy with the acquisition of DYNASIM. CATIA V6 SYSTEMS is based on 3 major axes: collaborative intelligence, fully supported by ENOVIA 3D LIVE, modelling and behavioural simulation, including DYMOLA technologies and the adoption of the MODELICA standard, and the SOA V5 infrastructure for collaborative lifecycle management, encompassing the capabilities of ENOVIA VPLM, SMARTEAM and MATRIXONE to support end-to-end processes, from concept to deployment.
Despite the functionalities that this modelling offers, there is no general model that covers all the properties. Hybrid or multiphysics models pose serious practical problems, particularly when it comes to calculating physical properties. These two observations lead to the need for conversion methods between the different representations of solids.

5.2- 1D physical modelling approaches and tools

5.2.1- General

One-dimensional (1D) physical modelling of multi-disciplinary systems is perfectly up-to-date. This is all the more the case as many problems are now arising with the advent of concurrent engineering and new technologies. However, our frame of reference is that of analytical modelling or Cartesian logic. Faced with a complex system, we are called upon to develop models on which to base our reasoning. The models used are generally analytical models. An analytical model can help guide the researcher towards acceptable solutions. However, we can criticise the advocates of the analytical school who, with their models, often try to model a multi-disciplinary system separately for each domain and adapt these models to the models available rather than looking for substitute models. Hence, to approach a multi-disciplinary system, it is necessary to change domain within the same system. In addition, the use of analytical models requires a simplification of the phenomenon being studied, and therefore a breakdown or division by phenomenon. This method has shown its limitations over time, while the modelling of today's complex systems is in search of sound methods.
The modelling and dynamic simulation of multidisciplinary systems has been a

subject of research for more than three decades. There are several approaches to modelling these systems. In general, these approaches can be divided into three basic categories:

Approaches based on integration at the formulation level ;

Approaches based on integration at equation level ;

Approaches based on integration at solution level.

We briefly outline these approaches in order to assess their relevance and applicability to the modelling of multidisciplinary systems.

5.2.2- *Integration at formulation level*

One of today's approaches to multidisciplinary modelling is to apply mathematical and physical formulations such as :

Bond graph theory [26] ;

Linear graph theory [27] ;

LAGRANGE formulation [28. ;

Constrained multibody formulation [29].

Bond graph theory

The bond graph tool is a graphical method for multidisciplinary modelling of dynamic systems. Its main advantage is to provide a link between the various components of a system, which vary greatly in nature and are often governed by non-linear equations that are sometimes difficult to model. This tool is both pedagogical and practical, highlighting the relationships between the flows and forces of systems combining mechanics, electricity, thermodynamics, heat, energy, etc., which makes it highly appreciated in mechatronics.

Bond graphs are also called graphes des liaisons in French, but it is preferable to keep the original term, as the graph des liaisons is a better-known concept for representing the graph of the links in a mechanical system. Modelling a system using a bond graph is fairly simple in the sense that there are few elements to set up and the procedure is relatively straightforward. The difficulty lies in having a good knowledge of physical systems, as is the case with fuzzy logic.

The bond graph method is based on two main principles: the graphical representation of power exchanges within a system and the analogy between variables from different physical domains.

The exchange of power between two elements A and B of a system is represented, as shown in the figure below, by a half-array (called a "link" or "bond") which carries two variables known as "power variables", generically called "effort" and "flux" (Figure 11), whose product ($e \times f$) represents the instantaneous power carried by this link. In an electrical circuit, these are the current (i) and the voltage (v).

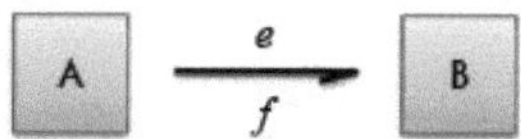

Figure 11. *Power exchange between A and B*

Linear graph theory

Like bond graph theory, line graph theory also represents the flow of energy through a system. The graph is made up of aretes that represent the flow of energy in a component, and arete boundaries that correspond to nodes. The energy flow is expressed by variables which are either forces and torques or speeds and displacements.

The graph is divided into chords and branches, and the chord and branch transformations are used to obtain the mechanical equilibrium, the kinematic loop and the closure equations respectively. The branch transformation uses the constrained multibody method. These transformations lead to the algebraic equations required to describe the mechanical part of the system. Similarly, to formulate the equations governing electrical systems, graph theory provides an approach for using the currents and voltages in the system. In this case, the transformations of the chords correspond to KIRCHHOFF's law of nodes, and the transformations of the branches correspond to the law of meshes. Line graph modelling is a fairly simple approach to modelling systems, and unlike bond graph modelling, it directly reflects the topology of the system.

Both approaches use KIRCHHOFF's laws of current and voltage to determine the equations governing the electrical system, techniques such as Lagrangian formulation, constrained multi-body formulation, and state-space modelling for the mechanical counterpart. These techniques can be used independently to model multidisciplinary systems.

Lagrangian formulation

The Lagrangian approach is also based on the analysis of the energy system. Mechanical systems are modelled in terms of variables that can be expressed as kinetic energy, dissipation energy and potential energy. The LAGRANGIAN formulation deals with scalar quantities, which facilitates its use when the number of elements in the system is minimal. The Lagrangian formulation has the disadvantage that its complexity increases exponentially as the number of components increases. The system of equations has to be derived individually for each element, and the differentiations become extremely cumbersome.

Examples of Lagrangian "*L*": *L = (kinetic energy - potential energy)* :

$$L = \frac{1}{2} m \dot{x}^2 - \frac{1}{2} k (x - x_0)^2 \qquad \textit{(Oscillateur harmonique)}$$

$$L = -m c^2/\gamma - q\phi + q\mathrm{A.v} \qquad \textit{(Charge q en électromagnétiques)}\ [\gamma = (1-\beta^2)^{-1/2}]$$

$$L \approx \frac{1}{2} m v^2 - q\phi + q\mathrm{A.v} - m c^2$$

$$L = \frac{1}{2} m (\dot{r}^2 + r^2 \dot{\theta}^2) + k/r \qquad \textit{(Loi de KEPLER ; Loi des orbites)}$$

$$L = \frac{1}{2} I_\perp(\dot{\theta}^2 + \dot{\phi}^2 \sin^2\theta) + \frac{1}{2} I_\parallel (\dot{\psi}^2 + \dot{\phi}\cos\theta)^2 - mgl\cos\theta$$

And to find the equations governing each system, we solve the following LAGRANGE equation:

$$\frac{d}{dt}\left(\frac{\partial L}{\partial \dot{q}_i}\right) - \left(\frac{\partial L}{\partial q_i}\right) = 0 \qquad i = 1, 2, \ldots n$$

With *qi* the generalized coordinates of each system *i*.

Constrained multi-body formulation

The constrained multibody formulation can be seen as a special case of the Lagrangian formulation, it overcomes the difficulties of the LAGRANGIAN formulation and can be effectively used to determine the governing equations of a multidisciplinary system. The basic idea is to use the maximum set of generalized coordinates for each body. We consider a general multibody system with *n* bodies and *m* constraints, each body has 6 degrees of freedom, i.e. the n-body system will have *6 x n* generalized coordinates. Consequently, the system will have 6 x n differential equations and n x m algebraic constraint equations. The constraint equations are given by :

$$\phi_j^i(q,t) = 0 \qquad i = 1, 2, \ldots, n \text{ et } j = 1, 2, \ldots, m$$

As for the Lagrangian equation to describe the system, it consists of 6 x n differential equations given by the equation :

$$M\ddot{q} - \phi_q^T \lambda = Q$$

Where: *Mis* the matrix of order (n x n), ϕ: is the constraint vector, λ: is the Lagrange multiplier vector and *q*: is the generalized force vector.

The two equations constitute a set of algebraic differential equations which can be solved using standard numerical methods. The electrical and control elements must be embedded in the above model by manually determining their appropriate equations.

The state-space representation is a well-known technique for the mathematical modelling of electrical and mechanical systems. This section briefly describes the general formulation. The EULER-LAGRANGE approach, NEWTON'S methods and KIRCHHOFF'S laws can be used to find the differential equations to describe the system. The general form of the state-space representation is as follows:

$$\dot{x}(t) = f\,(x,u,t) \qquad (1)$$

$$y(t) = g\,(x,u,t) \qquad (2)$$

In this order, (1) represents the state equation and (2) the output equation. In the case where these equations are linear around an operating point, they can be represented in the following form:

$$\dot{x}(t) = \mathrm{A}(t)\,x(t) + \mathrm{B}(t)\,u(t)$$

$$y(t) = \mathrm{C}(t)\,x(t) + \mathrm{D}(t)\,u(t)$$

Or: x(t) the state vector; y(t) the output vector; u(t) the input vector; A(t) the state matrix; B(t) the input (control) matrix; C(t) the output (measurement) matrix and D(t) the direct transfer matrix.

Bond graph theory uses the state-space approach to determine the equations governing the system, whereas linear graph theory uses the Lagrangian formulation with Kirchhoff's laws to define the equations of the system.

Although these methods have been used successfully for multi-domain modelling, they still have limitations in terms of their applications:

the formulations are not as a whole adapted to the components of the various domains of the model;

a lack of clarity in their scope of application for modelling new areas;

qualitative changes in system behaviour, which are common in electromechanical systems, cannot be managed by current approaches;

Current approaches do not provide integrated support for sensitivity analysis. Nevertheless, researchers have succeeded in applying this approach in certain areas.

5.2.3- ПёдгаМоп at liquation level

The second approach, which has become popular, can be classified as one that emphasises the integration between domains at the level of the system governing the equations. This approach gives an explicit mathematical formulation of the system governing the equation that can be used in sensitivity analysis. In the mid-1990s, two specific languages, Modelica and VHDL-AMS, were proposed for modelling multidisciplinary systems. The equation-level integration approach, the cornerstone of Modelica, is the language specification for multidisciplinary systems. Other integrated tools in this category include VHDL-AMS, OpenModelica and MathModelica. The following paragraphs briefly present each of these tools.

VHDL-AMS: VHDL-AMS is an analogue extension of the VHDL languages. It is a hardware description language based on a procedural paradigm. It was developed for the description and simulation of analogue, digital and mixed-signal systems, and is more oriented towards electrical/electronic domains. Several researchers have successfully used VHDL-AMS to model multidisciplinary systems. However, its simulation support and generalized library are not well developed.

MODELICA and its Environments - OPENMODELICA & MATHMODELICA: MODELICA is based on equation, an object-oriented language for modelling complex and heterogeneous physical systems. It was designed by the developers of DYMOLA, OMOLA, SIDOPS+, SMILE, OBJECTMATH and other modelling practitioners in various fields. This is why this language meets their specific needs. For other users who use SiMULiNK, cAMP-G, Adams etc.., This is one of the main reasons why the adoption of MODELICA's tools is not proposed in companies where engineers develop their models using SiMULiNK, ADAMS or others which are incompatible with MoDELicA. As a solution to this problem, interfaces such as SiMELicA (SiMULiNK and MoDELicA) are currently being developed, as is VHDL-AMS.

MODELICA is supported by a limited number of computing environments such as OPENMoDELiCA, DYMOLA, and MATHMODELICA. OPENMODELICA enables mathematical formulation and software architecture to be efficiently integrated using MODELICA. It has been developed for modelling and simulating complex, multidisciplinary systems. Its objective is to create a complete model as well as compile and simulate the environment. The tool generates explicit mathematical equations for multidisciplinary systems, but the compilation process is complex and involves lengthy procedures to generate the system equations. In addition, it requires the user to have in-depth knowledge of MoDELicA semantics, and the program compiler remains complex, with almost 100,000 lines of code. Also, MoDELicA determines a large number of equations which must be reduced intensively to reduce the computational effort.

MATHMoDELicA from MATHcoRE is also an integrated interactive development tool that provides a MoDELicA simulation environment that is tightly integrated with MATHEMATicA and MicRoSoFT ViSio. This environment has a graphical editor that is an extension of the MicRoSoFT ViSio diagramming tool, and symbolic manipulation is provided by MATHEMATicA. MATHMoDELicA is a commercial package, so its use for other research is limited. In addition, it does not provide support for design sensitivity analysis and parameter optimisation. It is important to note that MoDELicA and VHDL- AMS have specific languages and therefore require an external computational environment to actually simulate system response.

Based on the idea of equation-level integration, the researchers also used MATLAB symbolic tools (based on MAPLE) to create a system of equations modelled on multidisciplinary systems.

Despite its advantages, this approach has its limitations:

it does not support well the extension of parameters for the study of multidisciplinary systems;

requires the development of interfaces between software to overcome the incompatibility of languages;

the compilation process is complex and involves lengthy procedures to generate the system equations.

5.2.4- 1Мёдгаb'оп at solution level

Another strategy is to develop different software for the problems related to each domain, depending on the need, we integrate these software packages to model a multi-domain system, then we establish communication between them to simulate the behaviour of the system. Although this approach provides solutions in many cases, it is often cumbersome and does not produce an explicit mathematical formulation of the equations governing the system that can be used in sensitivity analysis. DYMOLA, DYNSAT and SIMPLORER are a few tools in this category.

SIMPLORER: Launched by ANSOFT CORPORATION, was introduced as a multi-domain simulation software in 2011. SIMPLORER is an integration of external simulators and provides a common platform for exchanging information between SIMULINK, MATLAB, MAXWELL, and sPICE simulators. The models are developed in C++, VHDL-AMs, FINITE ELEMENT ANALYsIs, RMXPRT, PEMAG, and use MATHCAD and MATLAB for mathematical manipulations.

SIMPLoRER was initially developed for transport and power applications. Later, it was extended to support other interfaces to meet the requirements of multidisciplinary systems. One of the main motivations for this evolution was the easy access to models provided by several suppliers over the Internet. This is one of the reasons why not all toolkits are well equipped to handle complex systems. For example, in the library

of SPICE SIMPLORER, the main drawback of semiconductor models is the basic electrical behaviour without taking dynamic effects into account.

DYNAST : DYNAST is a software package for modelling, simulating and analysing multidisciplinary dynamic and non-linear systems. It is capable of solving non-linear and non-stationary algebraic differential equations by performing analyses in time as well as in different domains. It uses an energy-based approach where, in each of the multipole models, the total energy interaction between a component and the rest of the dynamic system, assumes that interactions only take place in a limited number of energy component entries such as electrical connections, pipe entries, mechanical contacts, etc. DYNAST automatically formulates the equations describing the interactions between the components. Although DYNAST has been used successfully in many industrial problems such as electrical and magnetic circuits, mechanics, thermodynamics, and others, it does not support parametric optimisation well. Besides, it is a commercial software, in its free student edition it is limited to solve only 16 simultaneous equations.

DYMOLA : DYMOLA (DYNAMIC MODLING LABORATORY) is an environment for modelling and simulating integrated and complex systems. DYMOLA's multi-engineering allows the user to perform modelling and simulation on any physical component that can be described by ordinary algebraic differential equations. It is based on MoDELiCA, an object-oriented language for physical modelling developed by MoDELiCA ASSoCiATioN. The model components require development using MODELICA, and the set of differential and dynamic equations are then solved numerically using DYMoLA.

The limitations of this approach can be summarised as follows:

This approach does not provide an integrated formulation for modelling, simulation and design under a single platform.

Existing software is mainly developed for commercial use and the source code is not open to the public. As a result, while they work well as problem-solving tools, they do not provide the flexibility for further research.

All these software packages were designed for purposes other than modelling multidisciplinary systems. They have subsequently been modified to meet the current demand for modelling and simulation. This approach does not give an explicit mathematical formulation of the system governing the equation that can be used in analytical design.

5.3- The problem of multiphysics modelling

Methods and techniques for developing industrial systems have reached a certain maturity. Indeed, in recent years we have seen the emergence and application of approaches with associated tools dedicated to this subject. However, although these approaches provide a feasible solution in some cases, they remain unsuitable in many contexts. They have difficulty evolving to take account of new modes of design, modelling and simulation. In particular, they are often difficult to handle and have shown their limitations in the modelling of multidisciplinary systems. The evolving multi-disciplinary nature of these systems means that original solutions can be envisaged that have not hitherto been explored or exploited by technological advances.

In fact, most of the current system modelling processes deal with the trades separately. The drawback of these strategies is that the software they produce is not always satisfactory. Each simulation software is powerful in a particular field, but has limited skills in other fields. Hence the development of a large number of models specific to a field that is often dealt with (electronics, automation, mechanics, hydraulics, thermics, etc.). As a result, the simulation processes are limited to single-domain simulation and do not allow other simultaneous simulations of the various parameters required by industry (mechanical, hydraulic, electrical parameters, etc.). In this case, it is not possible to check the coherence

of its behaviour and easily optimise the variables of the system as a whole.
In addition, the classification of models by levels of abstraction hinders the simulation of multiphysical models. An approach based on the separation of abstraction levels, often considered as local, is well suited to the detailed analysis of the distribution of particular variables. In fact, it is reduced to the analysis of certain parts and is not adapted to the analysis of multi-domain physical and dynamic behaviour. In fact, the levels of abstraction of the models do not interact with each other to allow a complete analysis that takes into account the main phenomena of system operation and behaviour. We may therefore have difficulty in obtaining a desired representation of reality and visualising the behaviour of a multiphysics mechanism while analysing the system's behavioural parameters. As a result, these approaches do not provide an integrated formulation for design, modelling and simulation under a single platform for multidisciplinary systems.

6- Conclusion

Multi-disciplinary systems are a response to the demand for high-performance products that comply with new safety and efficiency standards. By their very nature, these systems encompass several disciplines at once (hydromechanics, electrothermics, electro-hydromechanics, etc.), making them increasingly complex. This evolving complexity has necessitated a change in design, from a trade-based and sequential methodology to a concurrent and complementary approach.
collaborative. Next, a review of the literature on system modelling and its sequential decomposition method highlighted the problem of modelling and simulating a multiphysical system. Indeed, the limitations of physical modelling approaches and the associated tools make 1D modelling difficult. In addition, the inability to communicate between the levels of abstraction of a multiphysics system model handicaps simultaneous simulation, which is advantageous in the development and analysis of such a system.
Indeed, particularly in the field of aeronautics, it is crucial to understand and evaluate the behaviour, performance, safety and other aspects of aircraft systems during their life cycle. In this context, we describe the multidisciplinary aspect of aeronautical systems before undertaking an in-depth multiphysics study of an aircraft system that is the subject of our work.

CHAPTER 3

SYSTEMS DESIGN MULTIPHYSICS AERONAUTICS

1- Introduction

Technological advances in aeronautics have always enabled the industry's equipment manufacturers to offer sound solutions for the civil and military aircraft markets.

In this chapter, we present a brief description of the main aeroplane systems which are multiphysical in nature. Firstly, we describe the hydraulic and electrical generation systems and the standard hydraulic components that make up an aeronautical multiphysics system. Next, we provide a functional and technical description of the main aircraft multiphysics systems. Emphasis will be placed on flight control and servo systems, which are almost identical on most aircraft (airliners or jetliners).

Finally, we are carrying out a detailed study of multiphysical aircraft braking systems. This multiphysics study encompasses the analysis of the system, the identification of the operating characteristics and the properties, both geometric and physical, of each of the components of these systems.

This chapter will therefore enable us to position our work by defining the problems associated with the modelling stage of a multiphysics aeronautical system.

2- Aircraft power generation

The operation of the various aircraft systems requires the generation of power from hydraulic and electrical sources (Figure 12).

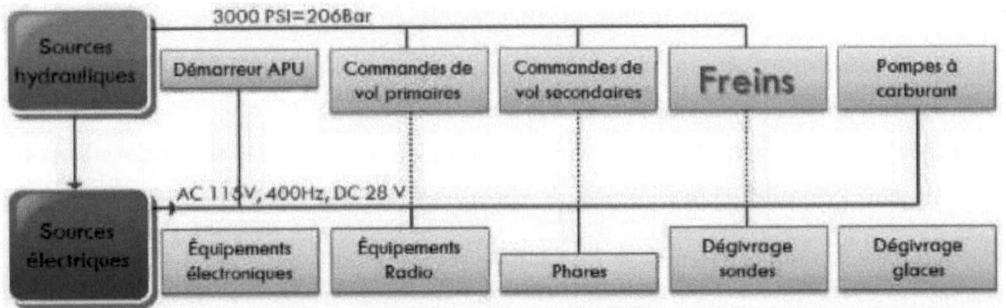

Figure 12. *Hydraulic and electrical generation of an aircraft*

2.1- Electric generation

The aircraft power supply system can be divided into three parts: an Alternating Current : AC"; a "Direct Current: DC"; and a "Standby" power supply provided by the battery (Figure 13).

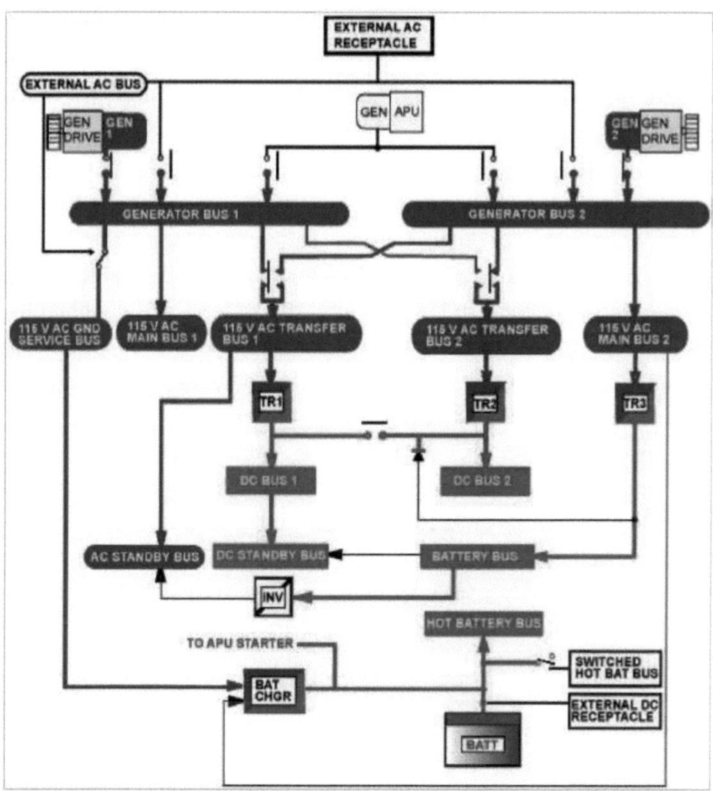

Figure 13. *Schematic diagram of the electrical generation of a Boeing 737 airliner* [30].

An aircraft's primary sources of electricity are usually two alternators driven by the engines to provide a three-phase alternating voltage of 115 volts, 400Hz.

Each alternator provides power to its bus network under normal conditions. Part of its network is transferred to the other alternator in the event of failure of the first. Transformers, on the other hand, provide low-voltage AC power for lighting and instruments that require this type of supply. Transformer rectifiers (TR) supply DC current.

The Auxiliary Power Unit : APU", for ground or emergency starting, is fitted with an alternator identical to that of the engines and can supply the two alternator buses (Generator bus) on the ground but a single bus in flight [31].

AC power supply: The AC power supply system consists of a generator bus, a main bus and a transfer bus.

In normal use, each engine alternator supplies its own alternator bus. If a fault occurs on one alternator bus, the associated transfer bus is automatically powered by the opposite alternator bus. Each transfer bus is associated with an

Alternate Transfer Relay (ATR) which automatically selects the opposite alternator bus. If the corresponding alternator fails and the transfer bus is set to AUTO, the APU can be used to replace a failed engine alternator. In this case the whole system will be powered normally.

The alternator bus supplies power to the servos and the main bus supplies non-essential servos. The transfer bus supplies essential services.

DC power supply: The DC power supply system provides power to devices requiring a 28V DC power supply (e.g. electro-hydraulic braking units). The three Transformer-Rectifiers convert 115V AC into 28V DC and are identified by TR1, TR2, TR3. Their output voltages vary between 24 and 30V. They are the primary sources of 28V DC. The battery provides power when no other source is available.

2.2- Hydraulic generation

Industrial hydraulic systems, particularly in aeronautics, require strength, flexibility and reliability. Today, the advent of electronics and information technology has made it possible to perfect hydraulic circuits. Since the early 50s, the complexity of flight control transmission systems, from the pilot to the control surface, and the stability problems of jet aircraft have led designers to use electrohydraulic commands to operate the control surfaces. The mechanical and aerodynamic forces due to the speed and large size of the aircraft make it impossible for a pilot to steer the control surfaces without hydraulic assistance. The same applies to braking, to generate mechanical friction at the wheels. Hydraulics are also needed for autopilots and aircraft with electric flight controls to make flying easier. Among the reasons for using hydraulic power on aircraft are [32] :

miniaturisation of hydraulic components ;

flexibility ;

high pressure easily achieved; acceptable operating temperature; operation independent of altitude.

Both fighter and airliner aircraft generally have two independent hydraulic circuits (Figure 14). The distribution is designed so that, in the event of the loss of one hydraulic circuit, all the systems continue to operate. In the event of the loss of two circuits, the main flight controls must continue to operate. In addition, in the event of a pipe rupture, numerous fuses isolate the defective part, thus preventing the loss of a complete circuit. In particular, the flight control servos are permanently powered by both circuits. The other systems are supplied normally by one circuit and in emergency by the other.

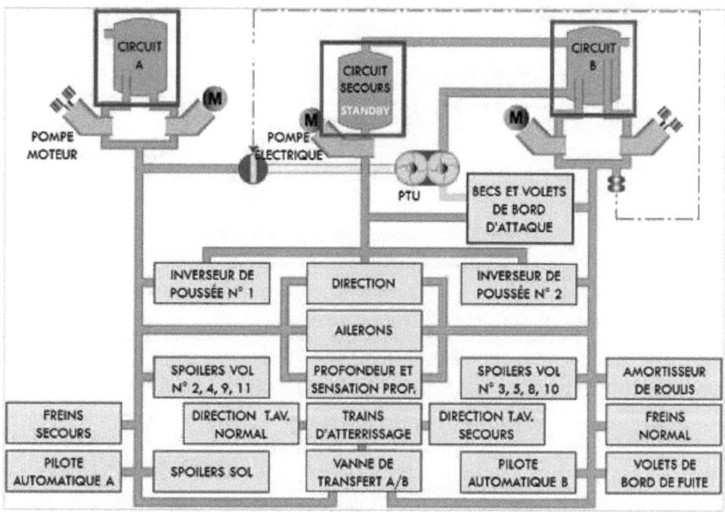

Figure 14. *Schematic diagram of hydraulic generation on an airliner* [33].

These hydraulic circuits operate at 3000psi, i.e. around 200bars. The pressure is supplied by hydraulic or hydroelectric pumps driven by the reactors. The Power Transfer Unit : PTU" is used to pressurise one circuit from another without transferring liquid.

A standard aeronautical hydraulic circuit essentially consists of a pump, a servo-control or distributor and a hydraulic receiver and other components: reservoirs, accumulators, filters, flow limiters, etc.

Hydroelectric servocontrols: Servocontrols are mainly controlled by a distributor that modulates hydraulic power. They are the most expensive components and the most difficult to design and manufacture. Servo-controls, generally hydraulic in aeronautics, are placed on the drive train to reduce the force on the control column to ensure smooth, precise control. They help the pilot to control the aircraft's speed and altitude with precision and little stick force (<0.25daN). Their contribution can be illustrated on an aircraft

The aircraft is equipped with a fighter jet, where the loads exerted by the control surfaces can reach 300daN. In these conditions, manual control would be impossible. Different types of servo can be used: hydromechanical; electrohydraulic; electromechanical. The servo control depends on the application, and is generally made up of [34] :

an input lever, transmitting commands from the pilot ;

a distributor that modulates power according to its opening;

a piston that converts hydraulic power into mechanical power;

a body and a rod to transmit mechanical forces;

mechanical links that provide servo-control of the servo's position (Figure 15).

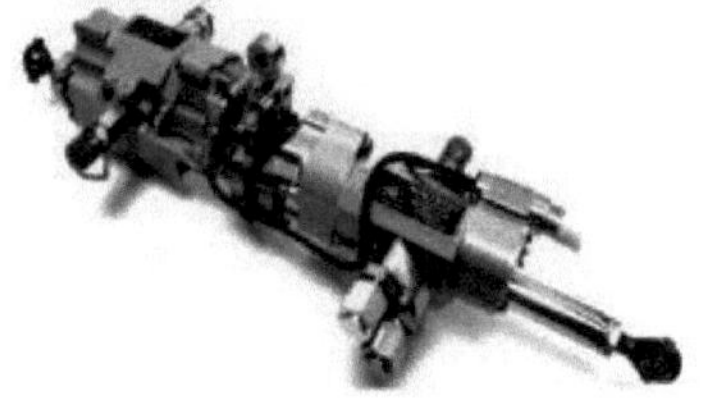

Figure 15. *Example of servo control*

Hydraulic receivers : Hydraulic receivers transform hydraulic energy into mechanical energy (Figure 16). We distinguish between: receivers for translational movement and receivers for rotational movement.

Figure 16. *Role of the hydraulic cylinder*

The cylinder is the energy-receiving element in a hydraulic circuit. It enables a very high force to be developed at a very precise speed. We distinguish between different types of cylinder depending on their design:

Single-acting cylinder : the rod-piston assembly moves in one direction only under the action of the pressurised fluid. Return is by spring or load;

Double-acting cylinder: the rod-piston assembly can move in both directions under the action of the fluid. The pushing force is slightly greater than the pulling force;

Cylinder with telescopic rod: single-acting for long strokes with a reasonable folded length.

3- Multidisciplinary aeronautical systems

In the aeronautics industry, on-board systems are designed to be increasingly high-performance in order to meet safety standards and the need for comfort and efficiency. This desire to guarantee safe and reliable avionics systems requires the development of new systems with enhanced performance and functionality [35]. With these advances in development in this field, the majority of aeronautical systems have become multiphysics systems (Figure 17).

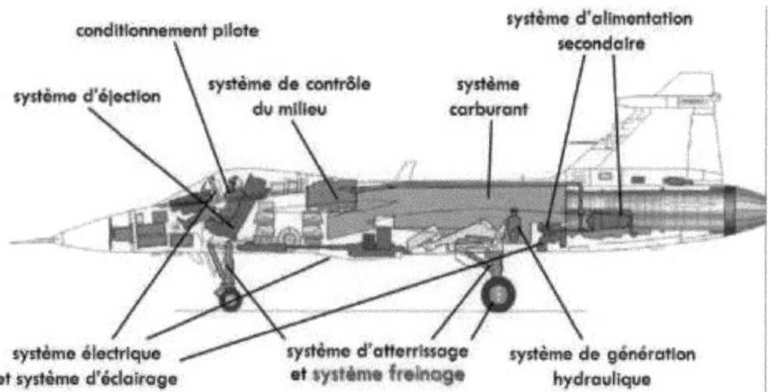

Figure 17. *Fighter aircraft systems*

An aircraft system/sub-system can be defined as "a combination of interdependent elements arranged to perform a specific function in an aircraft". The majority of aircraft systems are considered critical and are classified according to their functionality [36] :

Engine systems: these are the accessories that ensure the electrical, hydraulic and fuel generation of the aircraft engine. In addition to the engine, they include the ignition system, the fuel supply system and the control system;

Avionics systems: these are electronic and electrical in nature, and include navigation, communication, radar and weapons systems;

Cabin safety systems: these are comfort and assistance systems for people on board. They are designed in pneumatic, mechanical, thermal and electrical disciplines. They include air conditioning, pressurisation, oxygen, demisting, defrosting and ejector seat systems;

Flight control systems: are hydromechanical or electro-hydromechanical in nature and are used to operate the aircraft's control surfaces in order to control its trajectory and aerodynamic configuration around the aircraft's three axes (pitch, roll, yaw). They include the elevator, rudder, aileron and horizontal stabilizer control systems;

Servo systems: These are hydromechanical or electro- hydromechanical in nature. They help control the dynamic evolution of the aircraft on the ground. They include the landing gear, aerobrakes, nosewheel steering and aircraft braking systems.

The aircraft's manoeuvrability in flight and on the ground depends on the proper operation of the flight control and servo systems.

3.1- Flight control systems

Flight Control: CdV" systems are on-board command-control systems that steer the aircraft's trajectory by acting on its control surfaces in response to instructions given by the pilot. They provide the link between the pilot and the control surfaces, which are movable aerodynamic surfaces. These are all the elements between the cockpit* control units and the flight control surfaces designed to manage the aircraft's attitude, trajectory and speed in manual and automatic flight modes. In the cockpit, the pilot has control inputs (control column, rudder pedals) to influence the aircraft's attitude and trajectory. The pilot's input to the control column passes through a spring-loaded connecting rod fitted with artificial feel and decoupling devices to reach the servo-control for hydroelectric regulation of the control that operates the moving surface [37]. This architecture is described schematically in Figure 18.

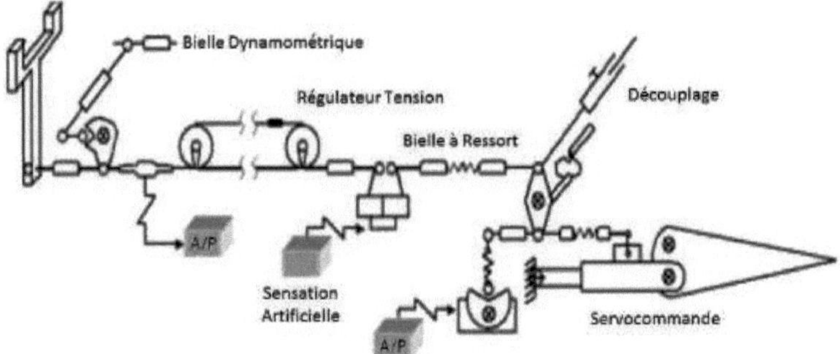

Figure 18. *Architecture of an airliner flight control system*

Fighter aircraft are equipped with flight control systems, which are either hydromechanical or electro-hydromechanical. They are used to operate the aircraft's control surfaces in order to control its trajectory and aerodynamic configuration around the aircraft's three axes (pitch, roll, yaw). They comprise the following systems

The cockpit of an aircraft is the space reserved for the pilot, co-pilot and mechanic. It contains all the controls and instruments needed to fly the aircraft, including elevator control, rudder control, aileron control and horizontal stabilizer control (Figure 19).

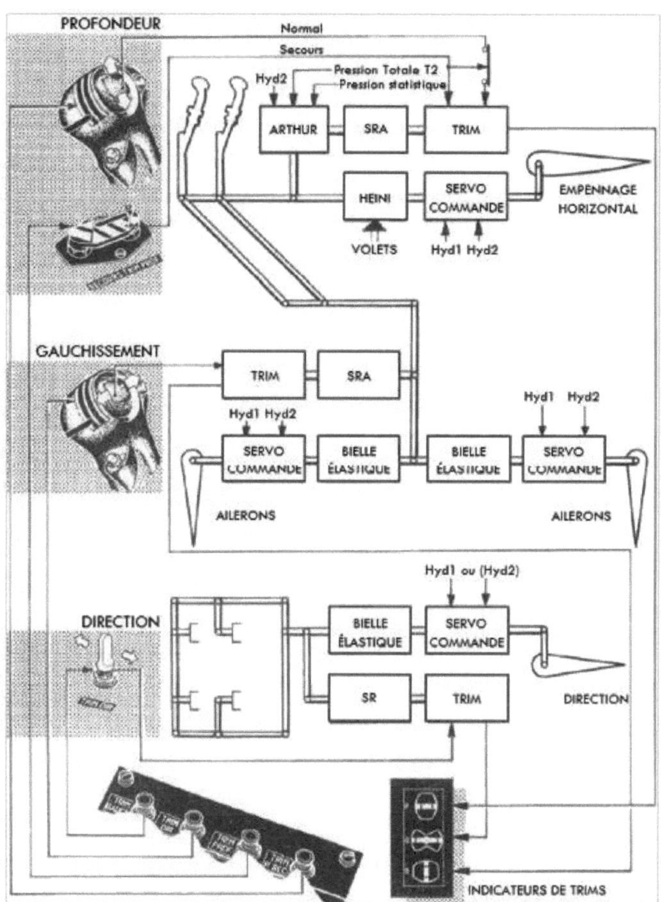

Figure 19. *Block diagram of the flight controls on a fighter aircraft* [38] Where: TRIM (Trim system); ARTHUR (Force proportionality system); SRA (Artificial Reaction system); HEINI (Gear reduction system).

The hydraulic servo-controls are assisted by electro-hydraulic devices: compensation system, artificial reaction system, force proportionality system, demultiplication system [39]. It is controlled by deflecting the one-piece horizontal stabilizer for pitch, two ailerons for roll and the rudder for yaw. These control surfaces are controlled by the servo-control system, which is activated by the pilot using the control stick. The control column triggers the elevator and pitch cam servos via a mechanical link. The rudder servo is controlled by the pilot's foot on the rudder pedal.

Depth chain, includes :

both legs (in the case of two-seater aircraft) ;

steering by rods and bellcranks;

an Arthur device which makes the force on the control column proportional to the speed of the aircraft and which is equipped with a failure signalling device;
an artificial reaction system and a trim cylinder;
a Heini reduction device: This device has a non-linear reduction ratio and compensates for the torque occurring at the output and input of the flaps;
a hydraulic servo that directly operates the horizontal stabilizer.
Warping chain, includes :
the two upper parts of the sleeve ;
steering by rods and bellcranks;
an artificial reaction system and a trim cylinder;
a spring-loaded connecting rod with a threshold on each aileron branch, which prevents the control linkage of the other aileron from jamming in the event of servo saturation;
a servo-control on each stick that operates the corresponding aileron.
Chame de Direction, includes :
the two spreaders ;
a linkage system using cables in the fuselage and rods and bellcranks in the drift;
an artificial reaction system and a trim cylinder;
an elastic connecting rod with a threshold which limits the pilot forces at the servo input;
a hydraulic servo that operates the rudder.
Artificial reaction systems are spring boxes with a triple slope on elevator and warp, and a double slope on rudder. Each spring box can be moved by the action of an electric actuator, which changes the balance position of the zero-effort control. On the depth chain, the Arthur device completes the SRA.

3.2- Servo systems

Aerofreins: The aerodynamic surfaces of aerofreins are located on the upper or lower part of the aircraft fuselage. They help to reduce aircraft speed by deploying two metal surfaces in the opposite direction to the wind. Each brake is actuated by a single-body hydraulic cylinder, supplied by the corresponding hydraulic circuit via an electro-distributor. It is held in the retracted position by a latch with mechanical locking and hydraulic release [40].
The pilot control is directed to the solenoid valve, which in turn authorises the hydraulic supply to the cylinders and hook-up boxes, and also ensures hydraulic locking of the cylinders in all positions other than the retracted position.
Landing gear: Normal operation of the landing gear
comprises three fully automatic hydroelectric sequences [41] :
hatch opening ;
manoeuvring the landing gear ;

closing the hatches (Figure 20).

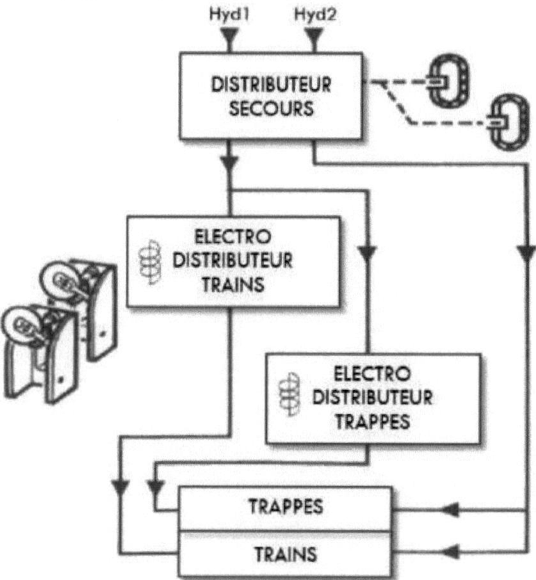

Figure 20: *Electrical and hydraulic exchanges in the* TRAPPES TRAINS TRAPPES *sequence* [42].

Hydraulic power is supplied by the corresponding hydraulic circuit. The undercarriage and trapdoor solenoid valves are supplied via an emergency valve. Each solenoid valve has an inlet and an outlet solenoid valve. The energised solenoid valve allows pressurised fluid to flow into one chamber of each piece of equipment, while the non-energised solenoid valve allows fluid from the other chamber to return to the hydraulic reservoir (Figure 21).

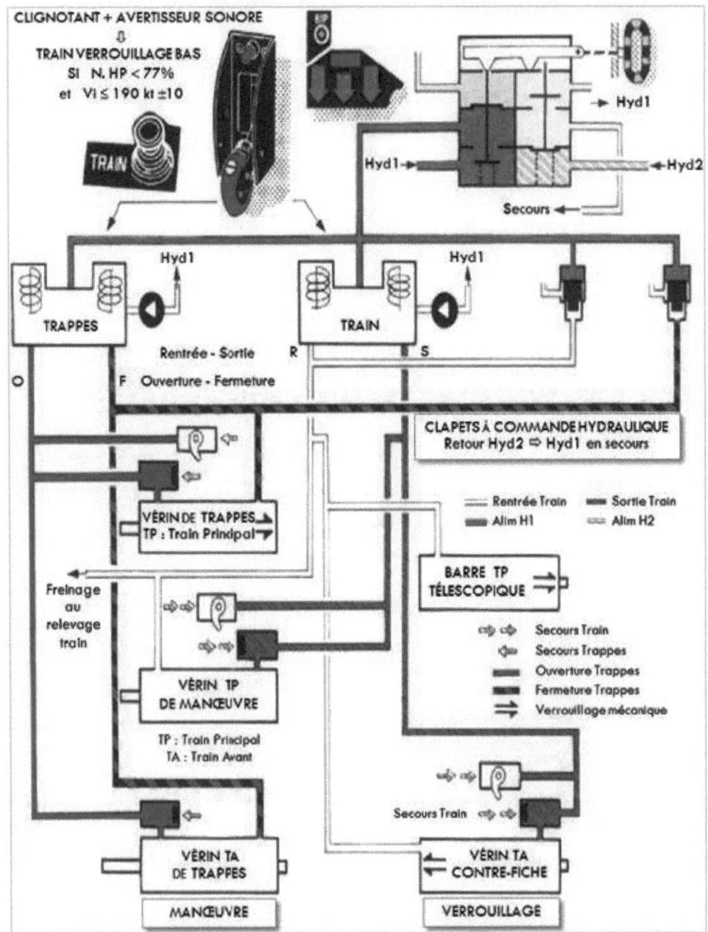

Figure 21. *Block diagram of normal operation of the Train circuit in a fighter aircraft* [43].

The electrical sequences are obtained by means of contactors activated by the end of travel of the doors and landing gear [44]. Some of these contactors also provide signalling [45]. For each sequence, the contactors are fitted in series. This means that if a contactor fails to operate, the entire manoeuvre is interrupted.

Each main lander comprises :

a landing gear leg ;

a manoeuvring cylinder ;

a telescopic bar which locks the landing gear leg at the bottom;

an attachment box that locks the landing gear leg at the top;

three doors, two of which are operated by a jack, the third being attached to the

landing gear leg;
a hatch lock that locks the hatch closed;
sequence and signalling contactors.
The nose gear comprises :
a landing gear leg with anti-shimmy system;
a counter-bolt for manoeuvring and locking the landing gear leg at the bottom;
an attachment box which locks the strut at the top;
three doors, two of which are operated by a jack, the third being attached to the landing gear;
sequence and signalling contactors.
Emergency operation is entirely hydromechanical and only allows the train to be extended. Hydraulic power is supplied by the second hydraulic generation circuit. The sequences are controlled manually. In emergency operation, the supply to the normal circuit is shut off at the emergency valve. The electrical circuit to the solenoid valves is rendered inoperative and the circuits for opening the hatches and lowering the landing gear are supplied directly by the second hydraulic generation circuit via the emergency valve. Two hydraulically-operated valves ensure the return of the return fluid to the first hydraulic generation circuit (Figure 22).

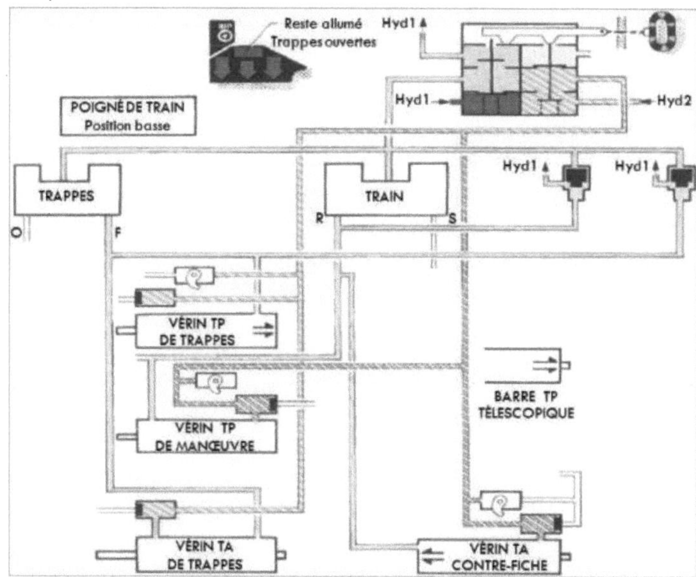

Figure 22. *Synoptic diagram of the emergency operation of the Train circuit of a hunting* [46]

In addition to the aerobrakes and landing gear, the aircraft braking system is part of this family of aircraft service systems. It is considered to be a critical system due to its significant influence on the dynamic behaviour and safety of the aircraft during taxiing. This criticality and influence increase for fighter aircraft, as malfunctioning of this system can have harmful effects on people, material assets and the environment.

Braking system: By design, aircraft braking is ensured by the operation of two independent systems [47] :

Mechanical braking: due to friction on the main wheels. This system is the most decisive in controlling the aircraft during the various stages of landing until the aircraft comes to a complete stop;

Aerodynamic braking: optional to reduce the landing speed, by deploying a parachute brake and/or extending the aerofoils.

Energetically, the moving aircraft has a kinetic energy proportional to its mass *m* and speed *v*, i.e. : $E_c =$ Я $m\, v^2$, and a potential energy proportional to its mass *m* and altitude *z*, i.e. : $E_p = m\, g\, z$.

A braking system is designed to be able to transform this mechanical energy into other energy. Most often, this recovered energy is dissipated in the form of heat either by gas compression (reduction in engine speed), or by the Joule effect (electric retarder) or by a mechanical friction system [48].

Figure 23 shows a block diagram of a multiphysics mechanical friction braking system for an airliner. Braking is electronically controlled and is actuated by hydraulic pressure from the Green circuit in normal operation. In the event of a failure in the Green circuit, an automatic switchover enables the Yellow emergency circuit to be used.

In the case of emergency braking, only the Yellow circuit is used in direct hydraulic connection with the brake pedals located on the rudder pedals. We can clearly see the importance of mechanics and hydraulics, and also the involvement of electricity and automatic control in such a system.

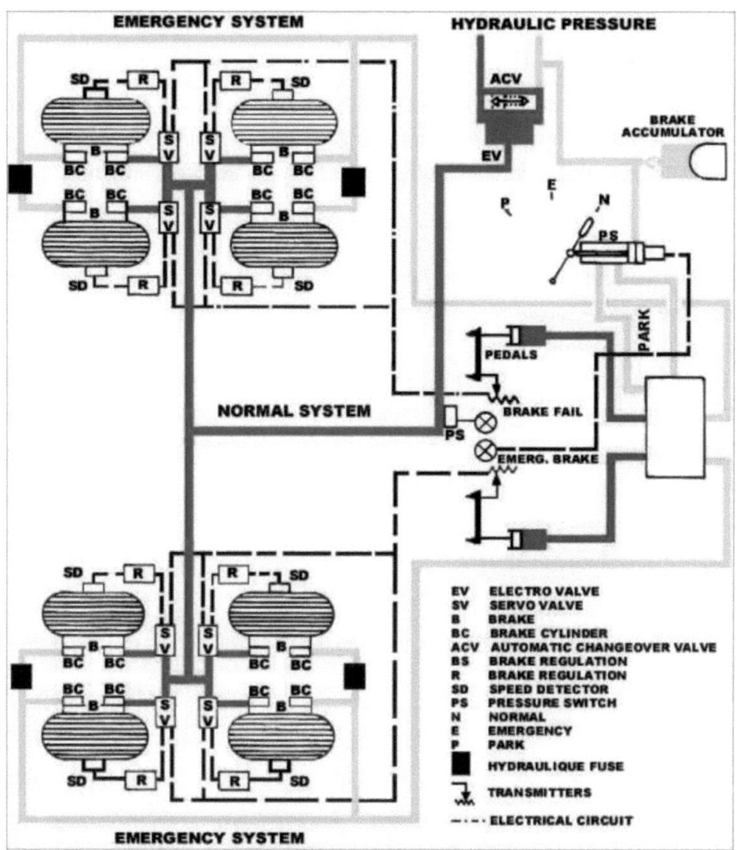

Figure 23. *Synoptic diagram of the multiphysics braking system on the Concorde aircraft*

For our study, we are interested in these multiphysical aircraft braking systems, especially for fighter aircraft, namely hydromechanical and electrohydraulic systems. They also qualify as critical systems, since they have a major influence on the dynamic behaviour of aircraft on the ground.

In order to model such a multidisciplinary aircraft braking system, we need to go through the system analysis and synthesis level [49] :

characterising the behaviour of the system: study of the operation of each component and of the complete system ;

define the system parameters and the constraints linking these parameters: these are the system specifications.

4- Study of a hydromechanical braking system

4.1- Hydraulic system supply

The hydraulic generation of a fighter aircraft is ensured by two identical and independent circuits:

utility system ;
flight control system [50].
These hydraulic circuits operate at 206bars. The pressure in each circuit is generated by a piston pump driven mechanically by the reactor.
The left engine drives the utility system pump, and the right engine drives the flight control system pump.
The flight control system provides hydraulic power to operate the flight controls. While the utility system generates hydraulic power to operate the servo systems: the landing gear, aerobrakes, nosewheel steering and the hydromechanical aircraft braking system.
This braking system operates at a constant hydraulic pressure of 83bars using a pressure reducer installed at the outlet of the utility hydraulic system (Figure 24).

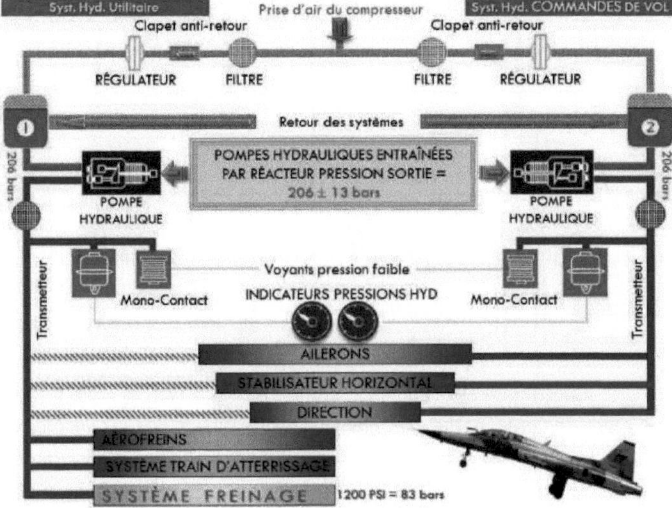

Figure 24. *Diagram of hydraulic generation on aircraft* 51]

The main flight controls (ailerons, rudder and horizontal stabilizer) are supplied by the two hydraulic systems in normal operation. However, if one of the systems fails, the hydraulic power required for these flight controls is supplied by the remaining system. The hydraulic reservoir of each system is pressurised by air drawn from the jet compressor. Consequently, the pneumatic discipline accentuates the multidisciplinarity of these systems (Figure 25).

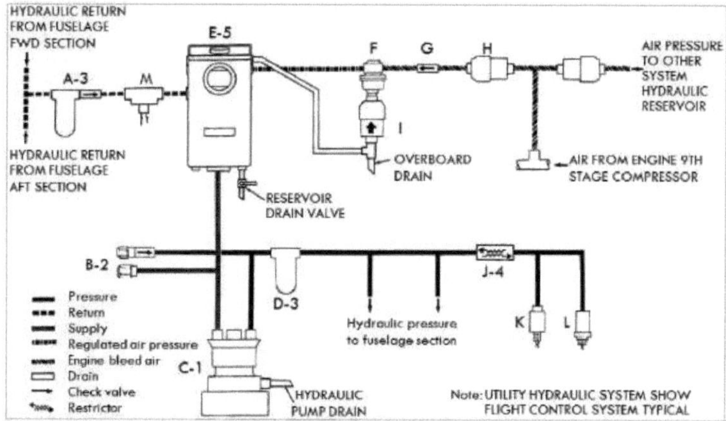

Figure 25. *Circuit of the fighter aircraft hydraulic generation system* * [51]

Legend for Figure 25 :

A Hydraulic filter to clean the return fluid before it enters the tank.

B Quick disconnect test sockets are used to connect to the hydraulic test bench for ground system testing.

C Hydraulic piston pump driven by the aircraft engine, system pressure delivered at 206 bar.

D Hydraulic filter that filters fluid from the hydraulic pump outlet and the test socket inlet when the hydraulic test bench is connected.

E Hydraulic tank holds 8 litres of hydraulic fluid and during normal operation at 1.1 ± 0.1 bar of air from the reactor.

The air pressure regulator maintains a pressure of 1.1 ± 0.1 bar in the hydraulic reservoir.

G Non-return valve prevents hydraulic oil or vapours from entering the air pressure line.

H hydraulic tank pressurisation air filter filters the air taken from the compressor.

i tank air filter which filters the exhaust air to the hydraulic tank when air compressor pressure is not available.

J damper-type choke valve installed in each system between the hydraulic system filter, pressure transducer, and reduces hydraulic flow pressure preventing surges and damage to the pressure transducer and pressure switch, while providing reliable gauge reading.

K Hydraulic pressure transmitter sends a pressure signal to the hydraulic pressure indicator.

The hydraulic pressure switch closes when the hydraulic pressure is reduced below 102 ± 6.8 bar, causing the hydraulic warning light in the cab to come on. The pressure switch opens when the hydraulic pressure is increased to 122.4 ± 6.8 bar, and the hydraulic warning light goes out.

M Hydraulic fluid temperature switch closes when the hydraulic temperature reaches 220 ± 5 F and opens when the temperature drops to 210 ± 10 F.

4.2- How the system works

Aircraft braking is provided by two identical, parallel-synchronised hydromechanical systems, each linked between the pilot pedal and the corresponding wheel. Each system consists of a mechanical control pedal, a hydromechanical brake distributor for brake pressure regulation and a brake block (Figure 26). The latter is made up of multiple discs mounted on the wheel of each leg of the corresponding main axle. The brake circuit is controlled by the physical force applied by the pilot's foot on the upper half of the rudder pedals and is supplied with pressure from the aircraft's utility hydraulic system.

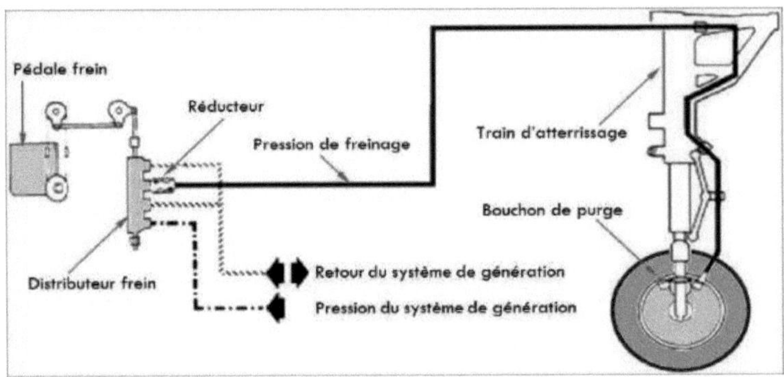

Figure 26. *Hydromechanical braking system for fighter aircraft* [46].

Utility hydraulic pressure is supplied to the brake distributor, which reproduces and regulates the required braking pressure according to a complex set of rules. This regulation includes proportionality to the force applied to the pedal, return pressure management and, above all, traction control. This results in hydraulic braking pressure that is directed towards the pistons of the wheel brake unit, which represent the hydraulic receiver of this circuit. This actuated receiver brings the stator discs and rotor discs into contact against a pressure plate fixed to the rim, producing mechanical braking friction. When the pedal is released or friction is reached, the braking pressure is reduced, following a reverse regulation at the distributor, and the piston return springs return the discs to their initial position.

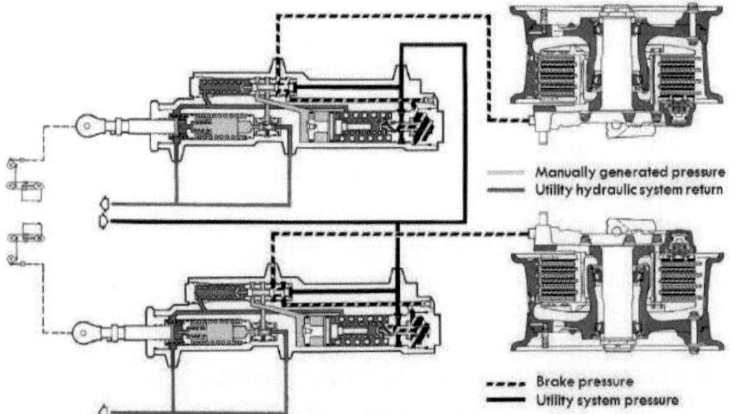

Figure 27. *Diagram of the hydromechanical blocks making up the brake system* [52].

Figure 27 illustrates the inlets and outlets of the brake system components and the flow diagram of the hydraulic flow. The directional control valve has an inlet port connected to the hydraulic generation source, an outlet port which supplies the brake block pistons with brake pressure, and two return ports.

The correct operation of the hydromechanical brake distributor is an essential and critical element in the efficiency and safety of this system. It is the pivotal component that joins the control element, the pedal mechanism anchored by the distributor rod, and the braking production element, the brake block, which is the brake pressure receiver. For efficient and coherent modelling, we have broken down our system into sub-systems.

The pedal is the braking control unit in the cabin. This control is purely mechanical, governed by a force law which regulates the relationship between the force exerted by the pilot's foot and the corresponding angle of angular translation (Figure 28). This law is given concrete form by a specific adjustment. In practice, the pedals can be adjusted separately using paddles located on the outside. Markings on the bulkheads of the front and rear stations enable the pedals to be adjusted synchronously.

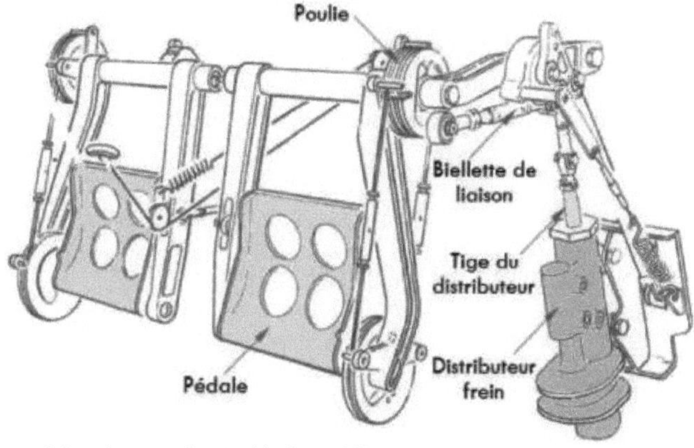

Figure 28. *Pedal and its mechanical linkage* 53]

The application of physical force on the pedals of the control lever generates an angular translation which is transformed through mechanical link poles into a linear translation through a link rod, which in turn results in a displacement of the brake valve spindle.

4.3- Hydromechanical brake distributor

The critical component of the aircraft braking system is the brake distributor, which modulates the hydraulic power of the hydraulic generator to generate the appropriate braking pressure to the brake block pistons. In addition, it receives brake pressure feedback once the desired friction has been reached. Thus, a complex reversible pressure regulation (Figure 29). Thus, the operation of this sub-system is decisive for the performance of the complete system.

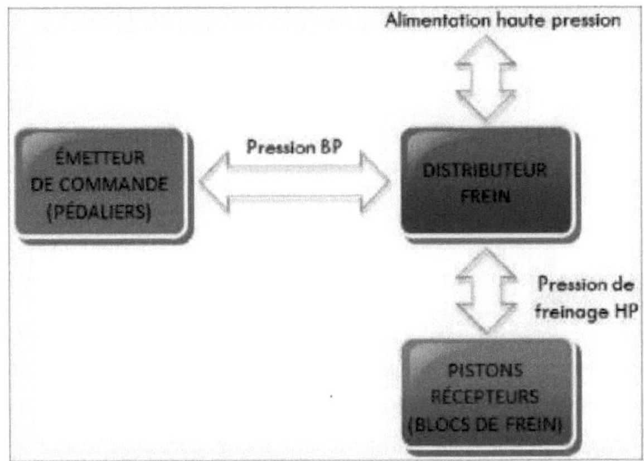

Figure 29. *Distributor role in aircraft braking*

This brake distributor consists of a connecting rod with pedal poles, a tulip valve, a master cylinder, a piston chamber, a metering valve, a pressure sensing valve, a shuttle valve and fluid flow interconnections (Figure 30).

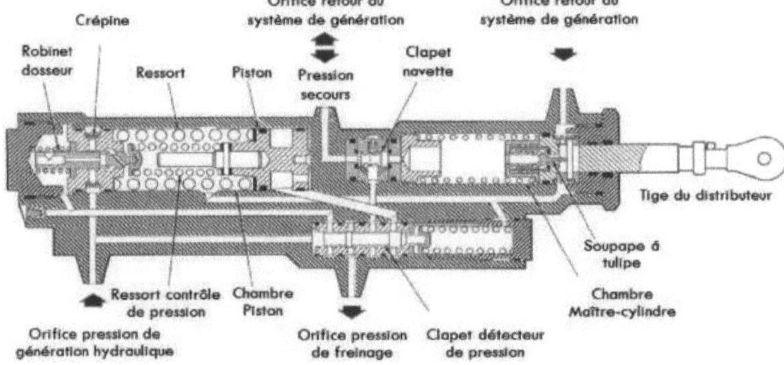

Figure 30. *Diagram of the brake distributor* [54].

At rest or in the brake released position, the return spring in the master cylinder chamber and the tulip valve maintain the maximum extension of the spool valve-pedal link (Figure 31). In fact, in the initial state, the pressure regulator, under the action of its spring, assumes a position which blocks the admission of hydraulic generation pressure. In this situation, the shuttle valve takes up the open position allowing communication between the master-cylinder chamber and the return of the generation system [55].

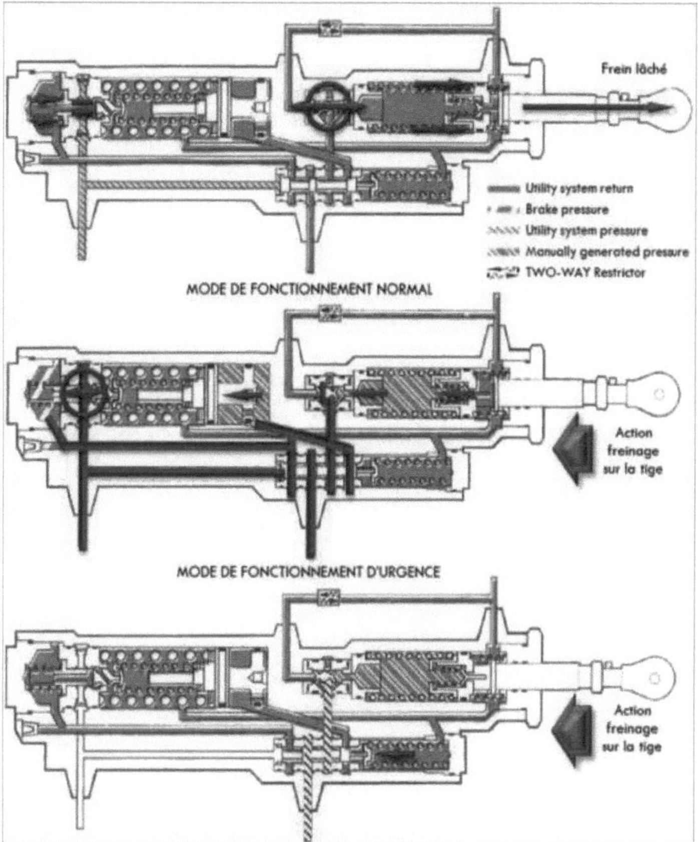

Figure 31. *Operating principle of the distributor* [56].

In normal operating mode, when there is pressure from the hydraulic system at the valve inlet, the application of an angular force on the upper half of the brake pedal results, via the connection of the poles, in a displacement of the valve stem which tends to press on the piston of the master-cylinder chamber and generates an increase in pressure which puts the shuttle valve in the closed position. This pressure causes the metering piston to move, exerting a force on the feeler spring. The movement of the piston causes the pressure metering valve spool to move through the small pressure control spring. The movement of the spool allows the brake pressure port and the hydraulic generation system pressure port to be aligned.

Similarly, the brake pressure proportional to the force exerted on the pedal is directed to the brake block. When the brake pressure fulfils the braking effort command, the orifice of the pressure proportioning valve spool closes (Figure 31). When the pedals are released, the spring action repositions the metering valve to

cut off the brake block's utility hydraulic pressure and allows hydraulic braking pressure to return by means of the four brake block piston return springs.
In emergency mode, in the event of a loss of pressure in the utility hydraulics system, the brake distributor operates like a brake master cylinder. As a result, applying greater force to the brake pedal causes the rod to move, manually generating pressure in the master cylinder chamber. This pressure is transmitted directly to the brake block through the pressure-sensing valve, making braking available and effective. Releasing the pedals causes the rod to return and the hydraulic fluid to return to the utility hydraulic system.

5 Study of an electrohydraulic braking system

5.1- Hydraulic system supply

The hydraulic generation of the braking system is also ensured by two independent circuits. In normal operation, each circuit is pressurised to 206bars. This pressure is delivered by a pump driven mechanically by the left-hand reactor for the left-hand circuit, and by the right-hand reactor for circuit 2. In emergency mode, an electrically-driven pump fitted as a bypass to circuit 2 compensates for the failure of the pump in the right-hand reactor (Figure 32).

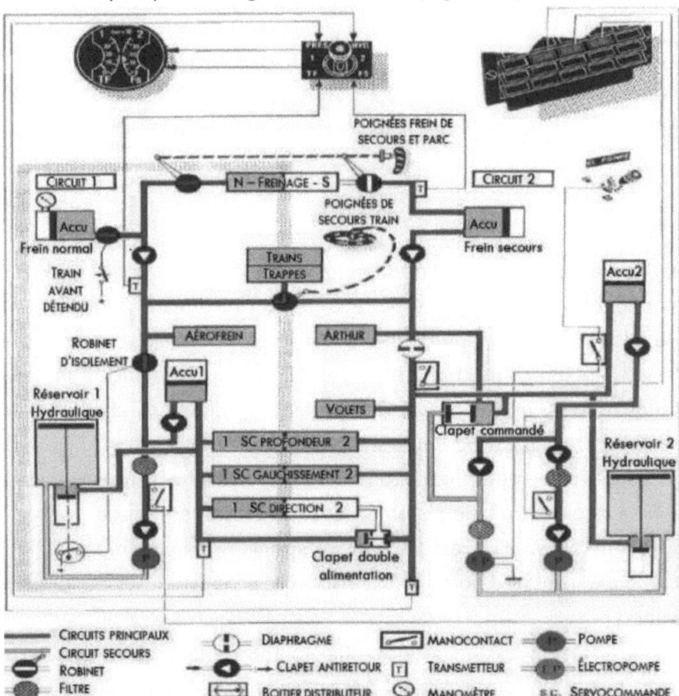

Figure 32. *Schematic diagram of the aircraft hydraulic generation system* [43].

Hydraulic circuit 1 supplies, via an accumulator, a body of elevator, pitch and rudder flight control servos. This hydraulic circuit also supplies, via an isolating valve, the servo systems: the aerobrakes circuit, the normal doors/trains circuit and the normal braking circuit.

In the event of a failure in hydraulic circuit 1, hydraulic circuit 2, via the normal accumulator, supplies the flight control servos, the high-lift flap actuator and also, via a control valve, the actuator of the Arthur force proportioning device, the emergency doors/trains circuit and the emergency braking/parking circuit.

For functional safety, hydraulic circuit 2 can be supplied by a secondary emergency supply, provided by hydraulic circuit 1, to take over in the event of a drop in pressure in this circuit. As soon as the pump in circuit 2 no longer provides the minimum pressure required, the constant flow electric pump (5l/min) starts up, enabling the components to remain operational. When the emergency circuit is pressurised by the electric pump, the control valve closes and the flow is directed as a priority to the flight controls and flaps, and through a diaphragm to the Arthur supply branch, emergency landing gear and emergency brake. The electric pump starts up when the pressure drops below 130bars and stops when the pressure reaches 175bars. A switch in the On position enables automatic operation of the electro-pump in the event of a failure in the hydraulic circuit2.

Hydraulic circuit 1 comprises :

a hydraulically pressurised tank with a capacity of 5 litres, equipped with a gauge and micro-switches;

a service isolation valve, which is automatically closed when the level in the tank falls below 2.1 litres;

a self-regulating pump with a flow rate of 23 l/min;

a filter ;

an accumulator for the servocontrols ;

an accumulator for normal braking ;

pressure switches and transmitters.

Hydraulic circuit 2 comprises :

a hydraulically pressurised tank with a capacity of 5 litres;

a self-regulating pump with a flow rate of 23 l/min;

a filter ;

a normal battery ;

an emergency brake battery ;

a double-feed valve ;

an emergency electric pump ;

transmitters and pressure switches.

From the above, the hydraulic generation of the normal braking system is ensured by hydraulic circuit 1 and that of the emergency braking system by the emergency hydraulic circuit of hydraulic circuit 2.

5.2- System circuits

The braking system is designed to absorb an energy corresponding to the distress energy leading to the destruction of the brakes after stopping. In standard conditions at sea level, this energy is achieved by stopping the 5300kg aircraft with a braking initiation speed of 119kt. Under the same conditions, the brakes absorb a normal amount of energy achieved, for example, by stopping the aircraft weighing 4500kg with a braking initiation speed of 100kt. To prevent the tyre from bursting after the brakes have heated up considerably, fuses built into the wheel melt under the effect of the heat and deflate the tyre.

The braking system comprises two independent circuits [42] :

a normal circuit: fitted with a traction control system, controlled by the front and rear pedals. Braking is differential and progressive;

an emergency circuit: mechanically controlled by the emergency brake lever on each station and used for automatic braking of the wheels when the landing gear is raised. The handle on the front station has a parking position.

5.2.1- Normal braking circuit

The normal braking circuit comprises two excitation circuits and a power circuit (Figure 33). The two hydraulic excitation circuits controlling normal braking are connected in such a way as to obtain differential and progressive braking, and consist of [57] :

four transmitters operated by the rudder pedals ;

two reservoirs supplying the transmitter circuit: a pressure relief valve evacuates the excess hydraulic fluid, via a common pipe, freely to the outside under the fuselage;

two independent pressure circuits regulated from the base plates, which are equipped with a by-pass, a solenoid valve and a non-return valve;

the power circuit consists of a hydroelectric expansion relay and an electronic traction control system.

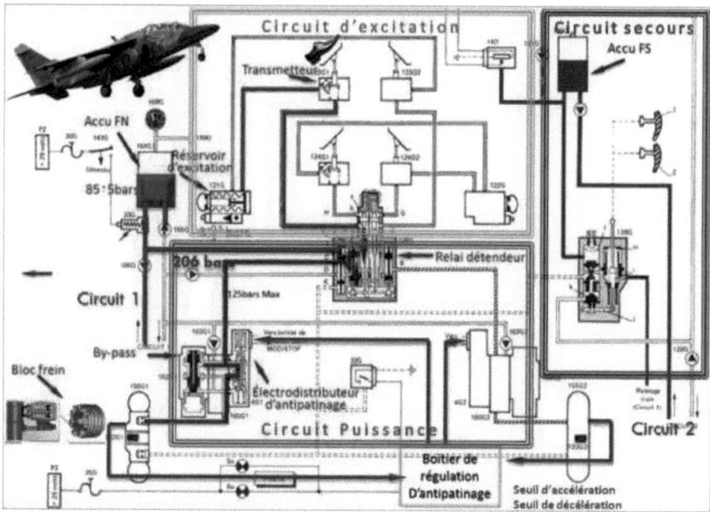

Figure 33. *Brake system circuits* [58].

In normal operation, tilting the pedals controls the excitation circuit which causes the transmitter piston to move, thus pressurising the excitation circuit and the chambers of the pressure reducing relay. This excitation pressure, acting on the normal pressure reducing relay, opens the supply orifice according to the position of the calibration spool. This opening allows the reduced pressure to pass to the base plate. This pressure delivered by the relay, relieved in proportion to the effort exerted on the pedal, is delivered to the brake blocks via the anti-skid system (Figure 34). As long as the pressure does not reach the bypass closing threshold, the brakes are supplied directly through the bypass. After that, it is directed to the solenoid valve which regulates the pressure supply to the brake blocks according to the information from the traction control system [59].

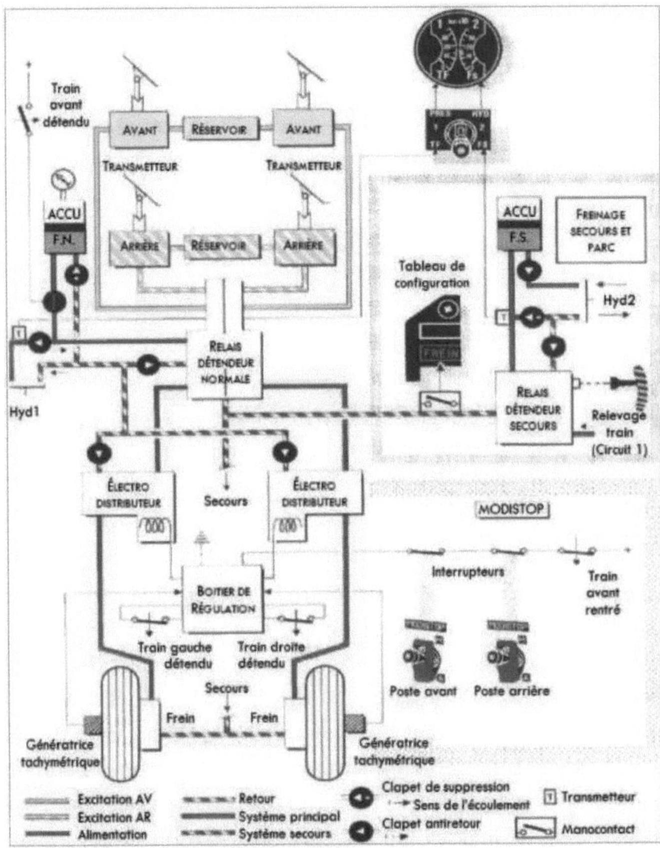

Figure 34. *Braking system circuits* [60].

As soon as the pedal is released, the pressure in the excitation circuit drops. The system is returned by its spring, and the relaxed pressure circuit communicates with the return circuit and masks the supply orifice. The drop in pressure causes the bypass to open.

The normal pressure reducing relay is supplied by hydraulic circuit 1 and by a brake accumulator isolated from the circuit during flight by a valve (microswitch). on the front shock absorber). Accumulator inflation pressure is around 85 ± 5 bar.

5.2.2- Emergency and park braking system

The emergency and pare braking circuit is controlled by the emergency brake handles, which are mechanically coupled in rotation and actuate an emergency pressure-reducing relay by means of a flexible control. The emergency pressure relief relay is supplied by hydraulic circuit 2 via an emergency brake accumulator. It simultaneously delivers pressure to both brakes, proportional to the rotation of the handles. This pressure is also directed to the normal pressure-relief relay in order to render the normal braking circuit inoperative. When the train retracts, the wheels are automatically braked by the emergency pressure reducing relay (Figure 35).

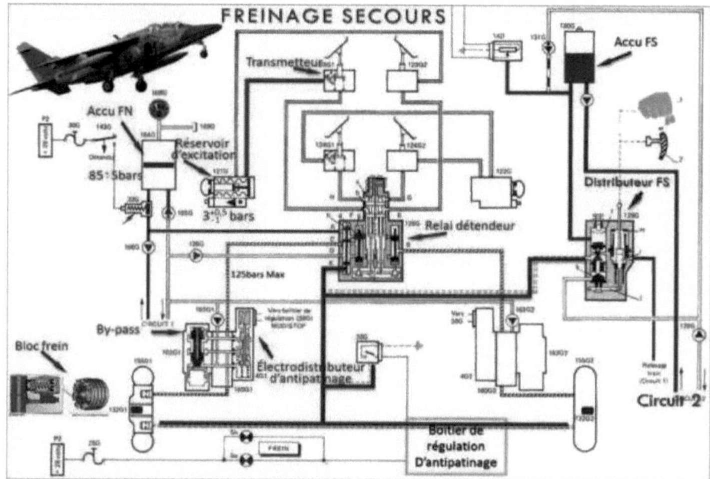

Figure 35. *Emergency braking operation* [61]

5.3- Traction control and brake blocks

5.3.1- Traction control sub-system

The traction control subsystem is a system that operates on the *on/off* principle (Figure 36). It comprises a control unit for each wheel [59] :

a solenoid valve mounted on the normal braking circuit;

a tachometric generator ;

a microswitch on the shock absorber.

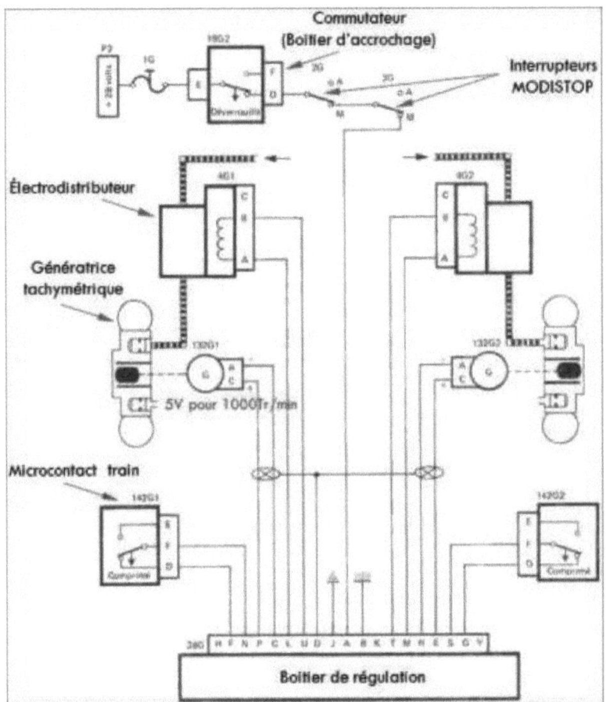

Figure 36. *Traction control subsystem* 62]

The traction control subsystem is powered by the P2 bar when the gear is down and the traction control switches at each station are set to On. From the

When the pressure in the brakes is too high and tends to lock the wheels, the control unit generates control signals for the solenoid valves that provide brake release. The control unit also generates the pre-impact braking prohibition, which is lifted by rotating the main wheels or compressing the main wheel shock absorbers. When activated by the control unit, the solenoid valve prevents the admission of pressure and causes this pressure to return to the circuit 1 reservoir. An anti-skid switch on each substation cuts the power supply to the system. If the switch is off, the charge on the capacitor in the control box drops to zero. When the traction control switch is reconnected, it takes a maximum of 2 seconds for this capacitor to recharge and give commands to the solenoid valves.

5.3.2- *Brake block sub-system*

The brake unit provides normal and emergency braking and includes a wear adjustment device (Figure 37). This equipment (weighing 15kg) enables the aircraft to be braked, immobilised when parked and steered when taxiing. It comprises [63] :

a hydraulic block ;

a torsion tube ;
a retaining plate ;
a FN connection ;
an FS connection ;
hot parts.

The brake block is operated by hydraulic pressure delivered on demand. This pressure acts on a set of four pistons fitted to the block to create mechanical braking friction.

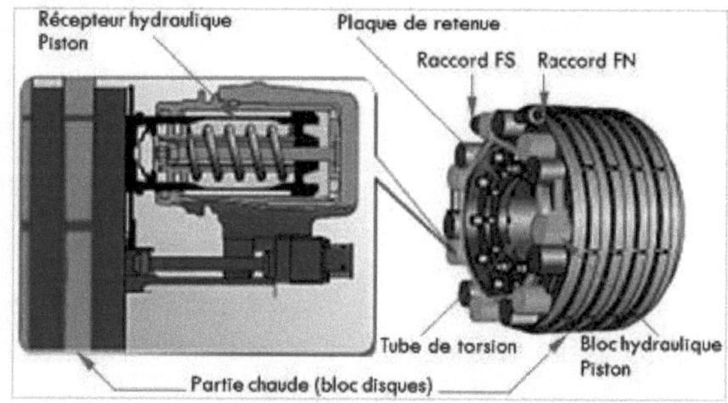

Figure 37. *Brake block*

6- Conclusion

In this chapter we have described the multiphysics aspect of aeronautical systems, focusing on the power sources of these systems, namely the electrical and hydraulic generations and also the standard components of aircraft hydraulic circuits. Next, we provide a brief inventory of the main aeronautical multiphysical systems. Finally, we presented a case study of two aircraft braking systems. We have illustrated the functional mechanical, hydraulic, electrical and electronic aspects of their sub-systems and components.

In order to study and analyse the behaviour of these multiphysical systems, a virtual prototype using modelling is required. To this end, we will present a versatile approach to the 1D physical modelling of multiphysical systems. This approach will be applied, using unified tools, to the two fighter aircraft braking systems that are the subject of our study.

CHAPTER 4

MULTIDISCIPLINARY APPROACH PHYSICAL MODELLING

1- Introduction

Today, the need to study the behaviour, validate the static and dynamic coherence and optimise the operation of new complex multiphysics products at all stages of their life cycles has led to the emergence of new organisational methods such as simultaneous engineering. The development of these products remains a costly process. As a result, new working methods based on the collaboration of several disciplines and drawing on modelling and simulation have become essential [64].

In this chapter we develop a method for 1D physical modelling of multidisciplinary systems.

The application concerns the modelling of two aircraft braking systems while exploiting the results obtained in the previous chapter. We illustrate how to model such a multi-disciplinary system using a piecemeal, component-oriented approach by undertaking unified modelling tools across physical disciplines.

2- Component-oriented modelling approach

The development of equations for the operation and behaviour of a multiphysical system is a task whose complexity increases considerably with the number and diversity of components involved in the system [65]. The usefulness of a multipurpose approach for multiphysical systems has already been demonstrated in numerous research studies.

The way to overcome this problem is to adopt a component-oriented modelling approach. This is an innovative multiphysics modelling method that consists of developing general models, instead of analytical models, that represent the system's components by means of specific values assigned to the physical and geometric parameters that characterise them.

In its implementation, this approach refers to systems engineering, which is based on the principle that a complex system is built by assembling sub-systems, in turn components. Hence, the design principle of this method is that large systems are designed from simpler systems [66].

2.1- Principle of the approach

The component-oriented approach is a cardinal principle for the modelling and analysis developed for different branches of engineering. It considers a system as a set of components constituting predefined libraries for different physical domains.

Component orientation enables the definition of modular architectures based on components. Each component is associated with a digital representation that can be reused and extended, making it useful in a wide range of applications, from modelling and simulation to implementation and control.

There are two types of component in a numerical model of a system: product components and process components. The former provide organic representations and the latter contribute to the cohesion of the model of the system to be modelled. Their digital prototypes support the evolution, adaptation and assembly of the components.

Managing the complexity of a system is based on three principles:

abstraction in the analysis of component behaviour independently of their internal representation;

breaking down complex systems into more basic subsystems and components ;

connecting sub-systems and components according to their interactions.

The composition of a model of a physics-based system is appropriately described in a one-dimensional geometry representation. The sub-models of a system must be linked together. To do this, each submodel is equipped with ports, which can have several inputs and outputs (Figure 38). Causality is enforced by linking the inputs of one submodel to the output of another submodel and vice versa. This approach makes it possible to simulate the behaviour of intelligent systems before their detailed CAD geometry is available.

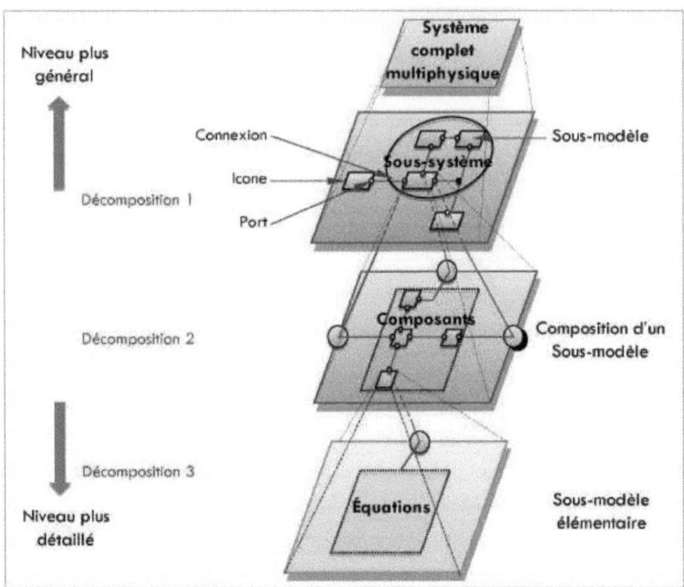

Figure 38. *Modular decomposition of a multiphysics system*

2.2- Features and benefits

2.2.1 - Characteristics of the approach

It is an approach to modelling multiphysical systems based on specific paradigms:

Causal modelling: The equations of each model are written in a declarative form, i.e. indicating the physical principles instead of writing the calculation algorithms, independently of the actual boundary conditions, and without deciding a priori which are the inputs and outputs. The result is causal models, described as DAE (Differential Algebraic Equations) systems, which represent the most natural and physically coherent way of representing each component of the system. The causality of the model is determined automatically by the simulation environment when a global model of the system is assembled from elementary models. In this way, models are easier to write, document and reuse; whereas the burden of determining the actual sequence of calculations required for simulation is left entirely to automatic manipulation tools;

Code transparency: the equations in a model are well matched to the way they are written on paper, so that it is easy to understand what is inside a given model, but also to modify or improve it;

Encapsulation: The interaction between components and 'sub' models is implemented by rigorously defined interfaces, called connectors, the design of which is of paramount importance. A connector is defined by a set of effort variables and a set of flow variables, while the simulation environment implements a connection by equating the effort variables and balancing the flow variables. For example: electrical connectors carry a voltage '*force*' and a current '*flow*', thermal connectors carry a temperature '*force*' and a heat flow '*flow*', drive train connectors carry an angle '*force*' and a torque '*flow*' and so on. As long as two different elements have compatible connectors, they can be linked together, regardless of their internal details (Table 1);

Heritage: models can have a hierarchical structure in which more complex models are obtained from simpler models.

Domain	Flow quantity	Level of effort
Electric	Current [A]	Voltage [V]
Translational mechanics	Force [N]	Speed [m/s]
Rotational mechanics	Torque [Nm]	Rotational speed [rad/s]
Hydraulics	Volume flow [m^3 /s]	Pressure [Pa]
Thermal	Heat flux [J/s]	Temperature [K]

Table 1. *Unified representation for different physical domains (P = effort xflux)*

2.2.2- Advantages of component-oriented modelling

In addition to the advantages of modular analysis and validation, reconfigurability, controllability and model interoperability, this approach offers the following benefits:

Modularity: modular programming allows code to be reused by writing ;

Reusability: Acausal modelling and encapsulation are strong incentives for the development of libraries of multi-purpose models that can be reused in different engineering domains;

Extensibility: This feature is essential for developing libraries of modifiable models. This makes it easy to replace part of a system model with a more detailed or simplified one, without affecting the rest of the model;

Multiphysics modelling: The scope of modelling is general. In fact, it provides modelling primitives such as general algebra, differential and differential equations, and is not linked to a particular field of engineering, such as mechanics, electrical engineering or thermodynamics. It is therefore relatively simple to use for modelling systems with a multidisciplinary character, such as mechatronic systems, resulting from the interaction of mechanical, electrical and control sub-systems, or nuclear power plants, resulting from the interaction of thermal-hydraulic, nuclear and control sub-systems.

Thus, by adopting such an approach, the complexity of the modelling process is limited to the development of elementary component equations and the standardised definition of component interfaces. Building a model by aggregating or 'connecting' the elementary components is both easy and intuitive, and generally results in a presentation similar to that of the physical system.

3- Implementing the approach

3.1- Software resources dedicated to the approach

Conventional modelling tools are usually used to simulate isolated physical domains: CATIA or Pro-Engineer for kinematics, ISIS for electronics, Fluent for fluid mechanics, etc. Multiphysics modelling tools are used to model and simulate systems involving several physical domains.

Today, the Simulink and AMESim tools have emerged as the standard for a modular system, as acausal modelling languages for complete and efficient modelling, and as simulation environments.

3.1.1- AMESim Imagine Lab

AMESim (Advanced Modeling Environment for performing Simulations) from LMS International is a simulation software for the modelling and analysis of 1D multi-domain systems. It is a language for modelling large, complex and heterogeneous physical systems, which is equipped with all the component-oriented modelling paradigms. It also provides an appropriate framework to support this component-oriented approach.

The software offers a 1D modelling and simulation suite for modelling and analysing multi-domain intelligent systems and predicting their multi-disciplinary performance. For example, mechatronic systems in robotics, automotive applications and aeronautics (with mechanical, electrical, hydraulic and control subsystems). It can be used to link sub-systems from different physical domains (hydraulics, pneumatics, mechanics, electrics, etc.). It includes a set of libraries of predefined one-dimensional geometric block models of components from different physical domains. Figure 39 shows an example of the block models.

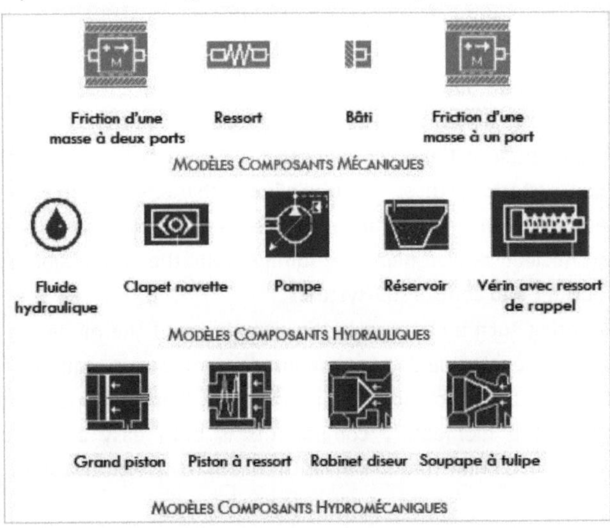

Figure 39. *AMEsim library components by domain*

Between the switch from submodel mode to parameter mode, the submodel is compiled. A compiler is therefore required. On the Windows platform, AMESim works with the free Gnu Compiler Collection (GCC) compiler, which is supplied with AMESim, but also with the Microsoft Visual C++ compiler.

3.1.2- Simscape / Simulink / MATLAB

The Simscape language is an extension of MATLAB's Simulink developed by MathWorks. It provides a 1D multiphysics modelling development environment with tools for modelling mechanical, electrical, hydraulic and other physical systems and domains. It was designed to create, develop and share 1D models of components (mechanical, electromechanical, hydraulic, etc.) in the Simulink environment. The Simscape software therefore facilitates the modelling, simulation and analysis of multi-domain physical systems. In addition, it makes it possible to design the physical model of a multi-disciplinary system without using the mathematical junctions of the components, while using the acausal approach of component-oriented modelling. It is therefore in the same family as the AMESim software. This type of tool is used for pre-dimensioning or validating system architecture before moving on to the deployment of detailed models on CAD tools.

Furthermore, unlike Simulink, where the wires between the blocks represented unidirectional flows of data (causality) and the blocks represented mathematical processing of this data, in Simscape the blocks represent physical elements such as a DC motor or a hydraulic cylinder, and the wires represent the mechanical shafts, electrical cables or hydraulic pipes that connect the elements. These wires model bidirectional data flows, the nature of which varies according to the physical domain. For example, in the electrical domain, the wires carry current and voltage data.

There are other 1D computer modelling tools that allow multiphysical systems to be modelled and therefore rendered functionally and aesthetically. Among the various physical modelling languages and tools developed from the end of the 70s, we can cite : ASCEND, OMOLA, gPROMS, MOSES, Modelica and Dymola. All these software packages were designed for a purpose other than the modelling of multidisciplinary systems. They have since been modified to meet current demand for modelling and simulation. They are based on object-oriented programming.

3.2- Language compatibility

Component-oriented modelling software is generally compatible with the object-oriented language. The object approach was invented to facilitate the evolution of complex applications. However, the difficulty with this type of modelling lies in creating an abstract representation, in the form of objects, of entities that have a material existence. This is the method that leads to software architectures based on the system's objects, rather than on the function it is supposed to perform.

3.2.1- *Principles of object-oriented programming*

Object-oriented programming is a computer programming paradigm developed by ALAN KAY in the 1970s. It consists of the definition and interaction of software building blocks called objects. Simula was the first programming language to implement the concept of classes in 1967. In 1976, Small talk implemented the concepts of encapsulation, aggregation and inheritance (the main concepts of the object approach).

The object-oriented approach considers software as a collection of separate objects, identified and possessing characteristics. A characteristic is either an attribute (data characterising the state of the object) or a behavioural entity of the object (a function). The functionality of the software then emerges from the interaction between the different objects that make it up. One of the distinctive features of this approach is that it brings together data and its associated processing within a single object.

Programming the objects of an application involves computer modelling of a set of elements from a part of the real world into a set of computer entities. These computer entities, which are called Objects, have the following characteristics:

the object data is quantified in discrete form;

every object has its own inherent identity;

two objects are distinct, even if all their attribute values are identical;

each object has a unique handle, by which it can be referred to, manipulated in various ways, such as an address, array index, or artificial number.

3.2.2- *Basic concepts of object-oriented programming*

The main basic concepts of this programming are :

Object: An object is an abstraction of an element in the real world. It possesses information, for example name, surname, address, etc., and behaves according to a set of operations that apply to it.

In addition, a set of attributes characterise the state of an object, and we have a set of operations (methods) that allow us to act on the behaviour of our object.

An object is the instance of a class, and a class is an abstract data type, characterised by properties (its attributes and methods) common to objects, which can be used to create objects with these properties.

Notion of class: A class is the structure of an object, i.e. the declaration of all the entities that make up an object. Objects of the same nature generally have the same structure and the same behaviour. The class factorises the common characteristics of these objects and enables them to be classified. An object is therefore an instantiation of a class, which is why we can speak indifferently of an object or an instance (or possibly an occurrence). All instances of a class are extensions of the class.

It is an abstract type of data characterised by properties (attributes and operations) common to its objects, and a mechanism for creating objects with these properties.

Class = instantiation + attributes (instance variables) + operations

Instantiation: The object has an identity that distinguishes it from other objects, regardless of its state. Instantiation represents the relationship between an object and its membership class, which was used to create it;

Attributes: They have a name and either a basic type (simple or constructed) or a class (the attribute refers to an object of the same or another class);

Operations: These are the operations (sometimes called methods) applicable to an object in the class. They can modify all or part of the state of an object and return values calculated from this state.

Notion of inheritance: Inheritance is a principle specific to object-oriented programming, allowing a new class to be created from an existing class. The name "heritage" (or sometimes class derivation) comes from the fact that the derived class (the newly created class) contains the attributes and methods of its superclass (the class from which it is derived).

The main advantage of inheritance is that we can define new attributes and new methods for the derived class, in addition to those inherited. In this way, we create a hierarchy of increasingly specialised classes.

Class hierarchy: Class hierarchy is an essential concept for data modelling. We can represent the parent relationship between different classes in the form of a class hierarchy, sometimes called a class tree. The hierarchy starts with a general class called superclass (base class, parent class, ancestor class, mother class or father class). The derived classes (daughter class or subclass) then become increasingly specialised;

Multiple inheritance: This technique allows a class to be inherited from two superclasses. In this way, the attributes and methods of several classes can be grouped together in a single class.

3.3- Objectives and method

The aim is to develop a multi-domain 1D physical model by considering a modelling approach that embraces the different disciplines of mechanics, hydraulics, hydromechanics, electricity, etc. and the following aspects: parametric modelling, reusability and interoperability of the models. To achieve this, a component-oriented modelling approach is adopted.

The physical modelling of the braking systems studied above will enable a virtual 1D physical model to be created from the real system, which will be validated by simulating the multiphysical behaviour of sub-systems and complete systems under the conditions of a functional test.

We adopt a fragmentary modelling of the components using a component-

oriented modelling environment. In addition, we create and develop our own models in C that can be integrated into the environment (lacking pre-modelled aeronautical components). Each model incorporates the data structure and behaviour of the corresponding component.
In this context, we have opted to work with high-performance software capable of offering one-dimensional multi-domain modelling that simultaneously takes into account the various physical phenomena involved in system behaviour.
We have opted for the Simscape / Simulink tool, which specialises in mechanics and hydraulics. On the other hand, we have opted for the AMESim tool, which deals with the majority of physics disciplines in a versatile way, particularly those of mechanics, hydraulics, electricity, electronics, thermics and pneumatics.
Firstly, we are developing the complete 1D model of the distributor using AMESim as the essential component of the hydromechanical system in order to validate it, and we are also producing 1D models using the same language of the complementary sub-systems of the entire system: hydraulic generation, pedal and brake block. In a second phase, we are developing the complete 1D model of the hydromechanical braking system fitted with an aircraft wheel using Simscape / Simulink / Matalb to simulate the friction force generated by the system at the wheel.
With the same objective of analysing the behaviour of the second electrohydraulic system, we are training 1D models in AMESim of the circuits making up this system. The interconnection of the 1D physical models of the sub-systems will make it possible to produce a 1D physical model for simulating and analysing aircraft braking systems.

3.3.1- Modelling steps in AMEsim

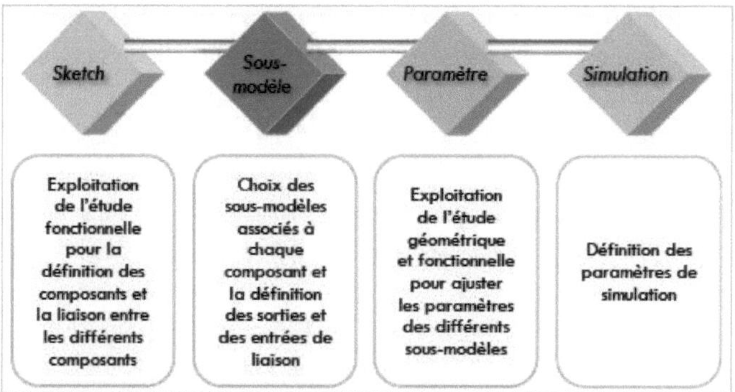

There are four stages in modelling a system:
Use of the functional study to define the components and the links between the various components
Definition of simulation parameters
Use of the geometric and functional study to adjust the parameters of the various sub-models
Choice of sub-models associated with each component and definition of link outputs and inputs

Figure 40: *Physical modelling steps in AMEsim*

Sketch mode: the library's components are implemented, rather than developed, following a detailed functional study of the system. In this way, we define the various components selected to represent the system's components and equipment;

Sub-model mode: during which we choose the physical sub-model associated with each component;

Parameter mode: during which we design the parameters for the different sub-models;

Simulation mode: when we run the simulation and generate the resulting curves for the parameters under study. (Figure 40).

3.3.2- Modëlisation stages in Simscape

Similarly, we build a model in Simscape in the manner of "What You See It's What You Get: WYSIWYG" software, i.e. the model is like a schematic representation of the system we want to simulate. For example, (Figure 54) shows a model of a "Resistance - Capacitance: RC" circuit with a generator and a voltage sensor. We can also see the *Solver Configuration* block, which is necessary for simulating Simscape models.

Simscape comes in the form of a set of libraries made up of multi-domain "physics" blocks and special functions. Simulink blocks provide the basic mathematical operations and Simscape blocks contain submodels of the basic physical components according to the physics discipline. Each submodel in each block has an internal structure and behaviour, and can communicate with its peers using the representation of its components and their relationships. The main Toolbox libraries are :

Foundation library: This is the basic SIMSCAPE library. It is the only one whose source code remains open. It is broken down into sub-libraries, one for each physical domain: Electrical, Hydraulic, Mechanical and Thermal component libraries;

SimDriveline : This library specialises in modelling mechanical power transmission systems;

SimElectronics: This library specialises in the modelling and simulation of electronic systems. It contains semi-conductors, logic circuits, integrated circuits, sensors, etc. This library also offers components compatible with the SPICE simulator;

SimMechanics: This library specialises in the modelling and simulation of mechanical systems. With this library, it is possible to simulate the kinematics of systems, taking into account the inertia of bodies and the constraints imposed by mechanical links;

SimPowerSystem: This library is not part of Simscape, but works on the same principle. It specialises in modelling and simulating power electronics systems (transformers, electrical machines, electrical networks, etc.). This library is not developed by Mathworks but by Hydro-Quebec, a company specialising in the production and transmission of electrical energy.

In each of the libraries, and for each domain, there are sources and sensors. For example, in the Foundation / Electrical library we find voltage and current generators and current and voltage sensors. In addition to these libraries, Simscape provides a solver and blocks for retrieving and processing Physical Signal data, which form the interface between SIMSCAPE sensors and generators and SIMULINK signals (under the Utilities library in Figure 41). However, the Toolbox / Simscape / Simulink libraries cannot contain all the component models. Hence the need to develop the missing models. Once the sub-models of the system components have been modelled and interconnected, the Simscape technology automatically constructs and generates the equations characterising the behaviour of each sub-system and integrates them with the rest of the sub-systems in order to establish the 1D physical model of the complete system.

Simscape 3.5 - The MathWorks, Inc. 2011

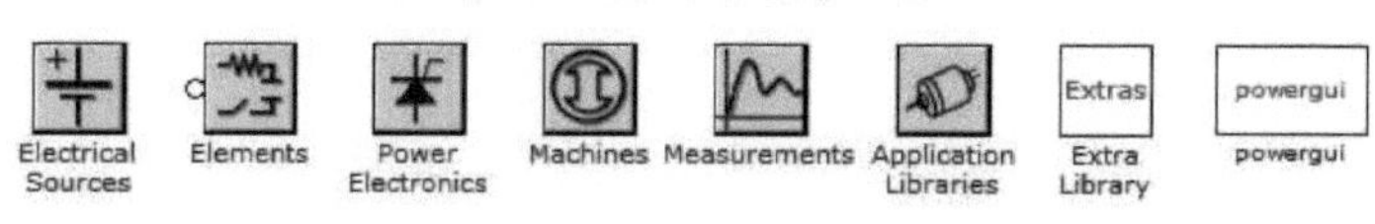

SimPowerSystems 5.4 - Hydro-Quebec and The MathWorks, Inc. 2011

Figure 41. *MATLAB 2011 - Toolbox Simscape 3.5 & Toolbox SimPowerSystem 5.4*

4- Modelling a dual-domain aeronautical system

4.1- Hydromechanical braking subsystem

The braking system of a fighter aircraft is ensured by the operation of hydromechanical braking friction applied to the main wheels. This friction takes place after hydraulic pressure has been released on the wheel brake block pistons. The reproduction and regulation of the braking pressure is supported by the complex hydromechanical brake distributor. This hydromechanical component

converts mechanical control of the pedal to hydraulic braking pressure at the wheel brake assembly.

4.1.1- 1D distributor model

The aim is to implement a numerical model of the distributor using AMESim, which is a dynamic simulation environment for engineering systems in various mechanical, electrical, hydraulic and thermal domains, as well as multi-domain physical networks. In fact, models of components from these different domains and their combinations can be harmoniously combined in this interface.

The study of such a multidisciplinary system requires the identification of the operating characteristics of each component and focuses on discovering the operating principle and geometric properties of the brake system. This study helps to develop a 1D physical model of the system in the AMESim language by adopting component-oriented modelling. The use of the functional and physical study, as well as the dimensioning, enabled us first of all to design the 3D geometric model of the aircraft brake distributor illustrated in (Figure 42).

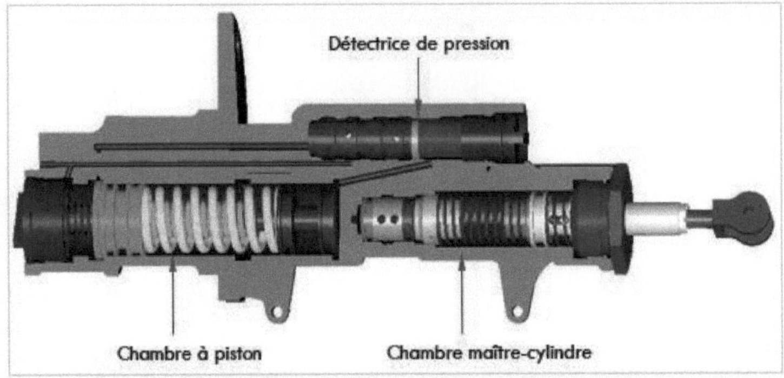

Figure 42. *3D geometric model of the aircraft brake distributor*

There are three main hydromechanical chambers:
pressure sensing chamber ;
master cylinder chamber ;
piston chamber.

Geometric and physical characteristics of the dispenser :

The geometric properties of the components (dimensions of the various components and spring stiffnesses of the spool valve) and the physical properties of the spool valve components (materials: Duralumin, Young's modulus: 74000MPa, Poisson's ratio: 0.33) are described in (Appendix 1).

We proceed to create the physical model of the distributor by adapting the following assumptions:

System input data: The pressure delivered by the hydraulic generation system is constant at approximately 83bars, and is delivered directly to the distributor via the pressure input port;

the complexity of the components and operation of the dispenser leads us to make a number of assumptions and simplifications about the operating mode and some of the model's components;

consideration of normal operating mode (regulated hydraulic generation pressure) ;

hydraulic fluid: specific fluid used in aeronautics.

We show the steps involved in modelling a dispenser chamber in the AMEsim environment, with reference to the results of the functional study, as outlined above.

Modeling the master cylinder chamber

First step: *Sketch mode*

In this first phase, we use the detailed study of the dimensional and functional aspects of the master cylinder's components to select the corresponding library components and implement them in sketch mode.

In order to distinguish the different physical areas involved in the operation of our multi-disciplinary dispenser, the components are colour-coded: mechanics 'green', fluid 'blue', hydraulic super-elements 'dark red' and signals, controls and observers 'red' (Figure 43).

Figure 43: *Implementation of master cylinder chamber components*

Second stage: *Sub-model mode*

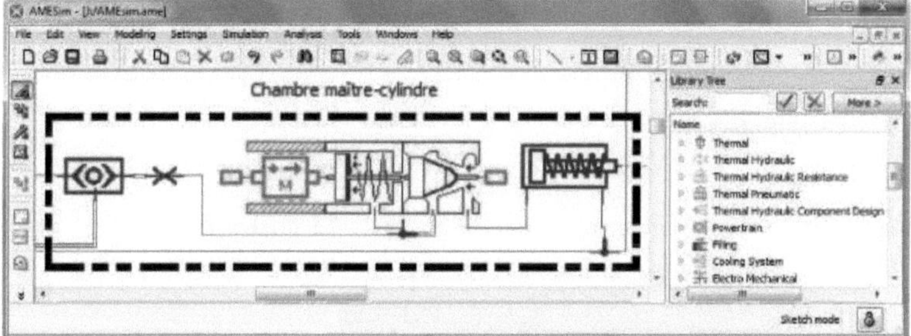

Figure 44. *Exchange links between the components of the master cylinder chamber*

In the second stage, the components making up the master-cylinder chamber are linked by hydraulic flow exchange or force transfer lines to form a single sub-assembly (Figure 44).

Third step: *Parameter mode*

In the Parameters window, we set the parameters for each component of the submodel, adjusting the geometric parameters: length, thickness, diameter, etc. and the numerical parameters: stiffness, initial displacement, etc. (Figure 45).

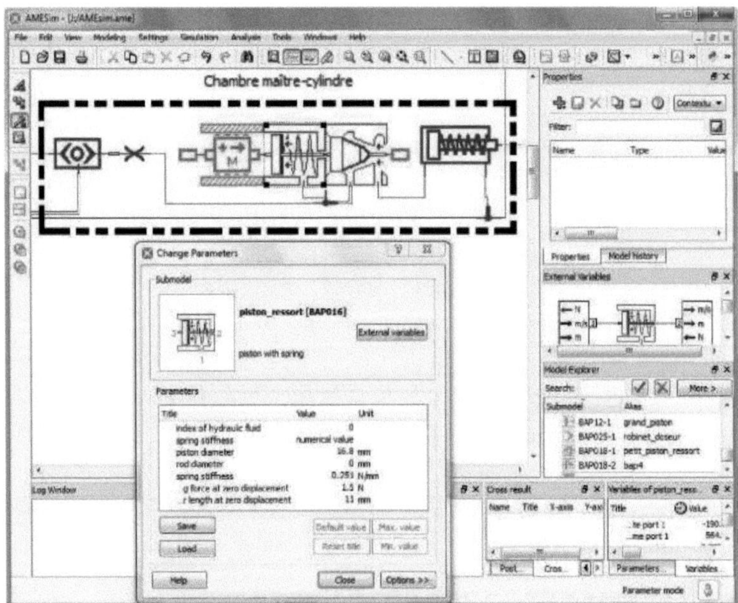

Figure 45. *Master cylinder chamber component settings*

Thus we obtain the 1D physical model under AMEsim illustrated in (Figure 46) :

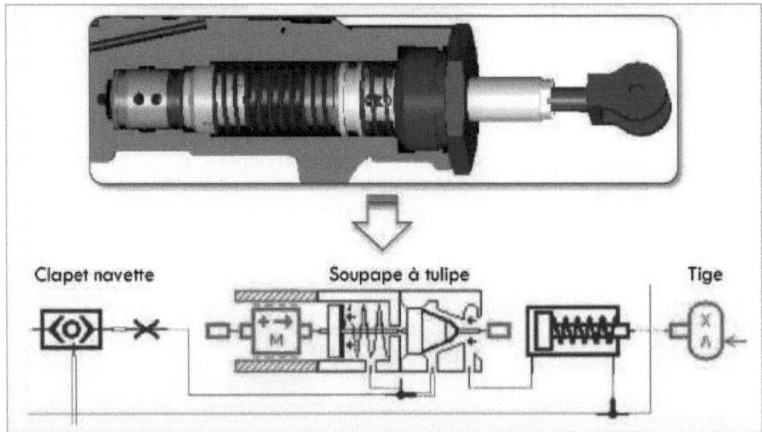

Figure 46. *AMEsim model of the master cylinder chamber*

Modeling the piston chamber

In the same way as the master-cylinder chamber, we are developing the 1D physical model of the piston chamber (Figure 47):

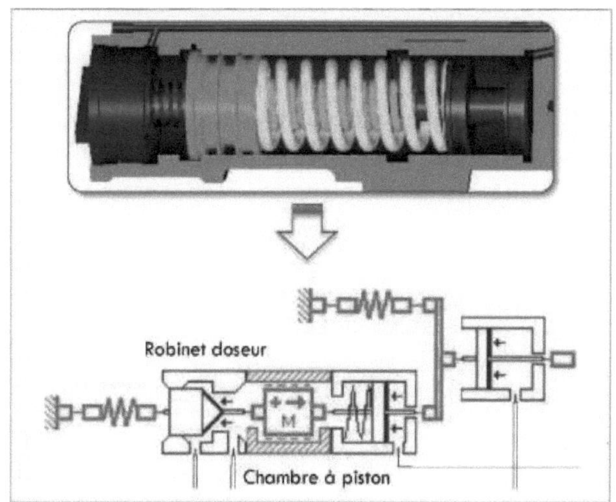

Figure 47. *AMEsim model of the piston chamber*

By interconnecting the chambers that make up the distributor, we can obtain the AMEsim model of the aircraft brake distributor (Figure 48):

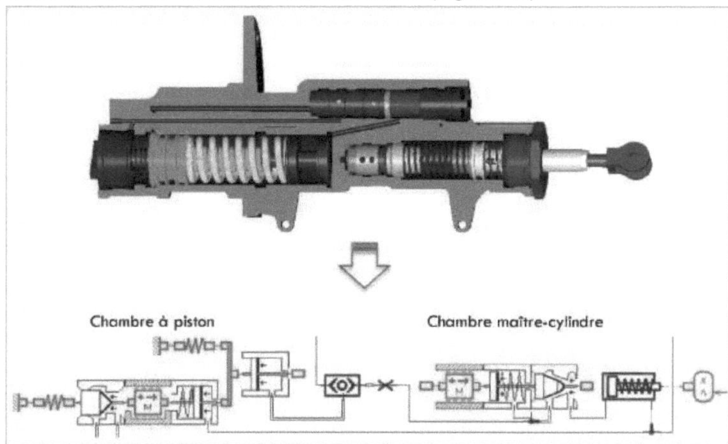

Figure 48. *AMEsim model of the brake distributor*

4.1.2- 1D model of brake subsystems

Based on a study of the functional and geometric architectures, we have defined the physical and dimensional properties of the brake subsystems: aircraft hydraulic generation, aircraft pedal, brake block. We then developed the 1D physical models, using AMESim, of these complementary sub-systems of the aircraft braking system, adapting the following assumptions:

hydraulic generation system: generates a constant pressure of 83 bar;

modelling the inputs to the hydraulic generation system and the pilot pedal using signals ;

modelling the brake unit using a piston to receive the hydraulic braking pressure generated by the distributor ;

the system is controlled by the force exerted on the pedal;

hydraulic fluid: specific fluid used in aeronautics.

Modeling the aircraft pedal

We model the pedal using the *Vehicle dynamics* library and control the braking action using an input signal and a signal-to-force converter representing the braking force applied by the driver's foot (Figure 49):

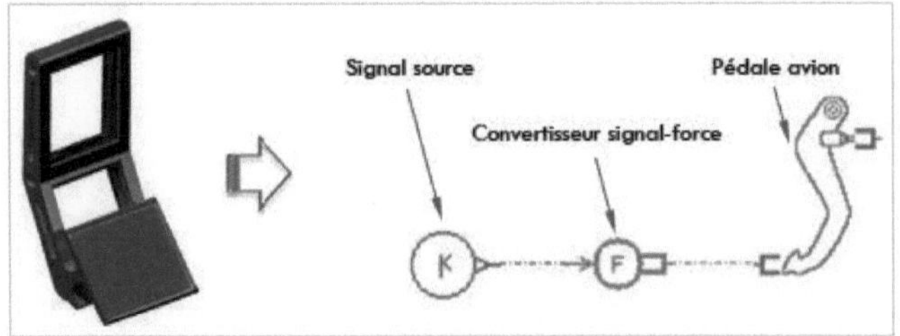

Figure 49: *AMEsim model of pedal and brake control*

Brake block modelling

The braking pressure developed by the distributor in response to the braking command is converted by the brake block pistons, which generate braking friction between the rotor and stator discs:

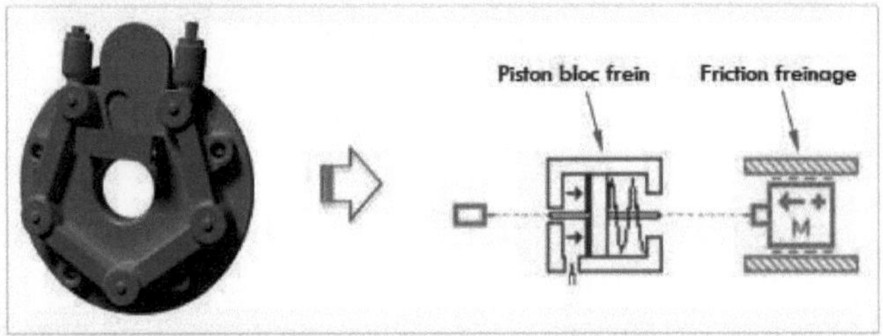

Figure 50: *AMEsim model of a brake block piston and braking friction*

Tank and hydraulic fluid modelling

A study of the characteristics of the hydraulic fluid used in aeronautics has enabled us to define the parameters to be inserted in the fluid data sheet (Figure 51):

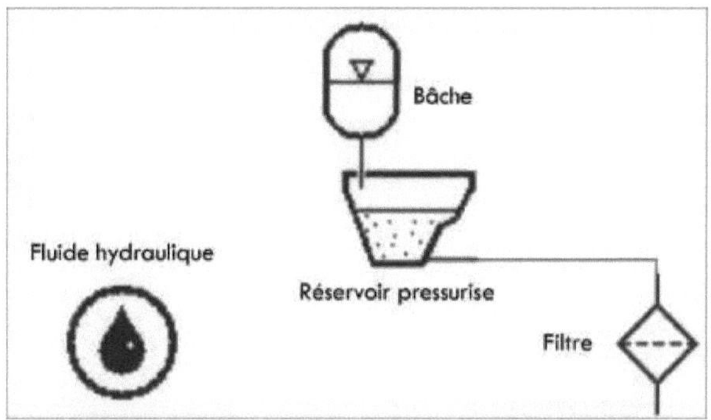

Figure 51. *AMEsim model of hydraulic reservoir and fluid*

Modelling the hydraulic generation system

The pressure of the hydraulic generation system is modelled to simulate the constantly delivered pressure P_h = 83bars. The AMESim model of the aircraft hydraulic generation system is shown in (Figure 52):

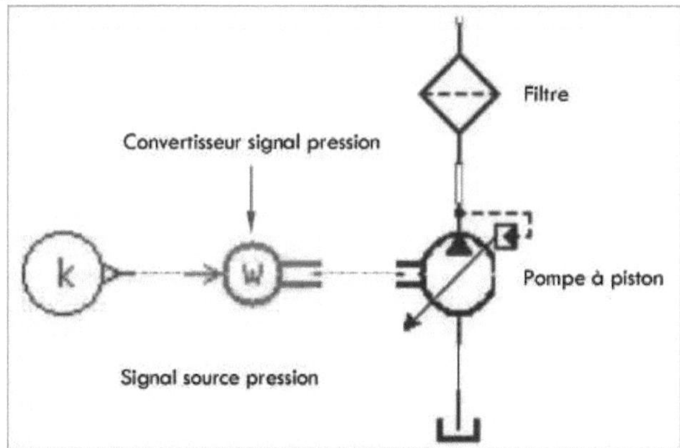

Figure 52. *AMEsim model of hydraulic generation system*

4.2- Hydromechanical braking system

4.2.1 - AMESim system model

The interconnection of the sub-system models results in the 1D model of the system

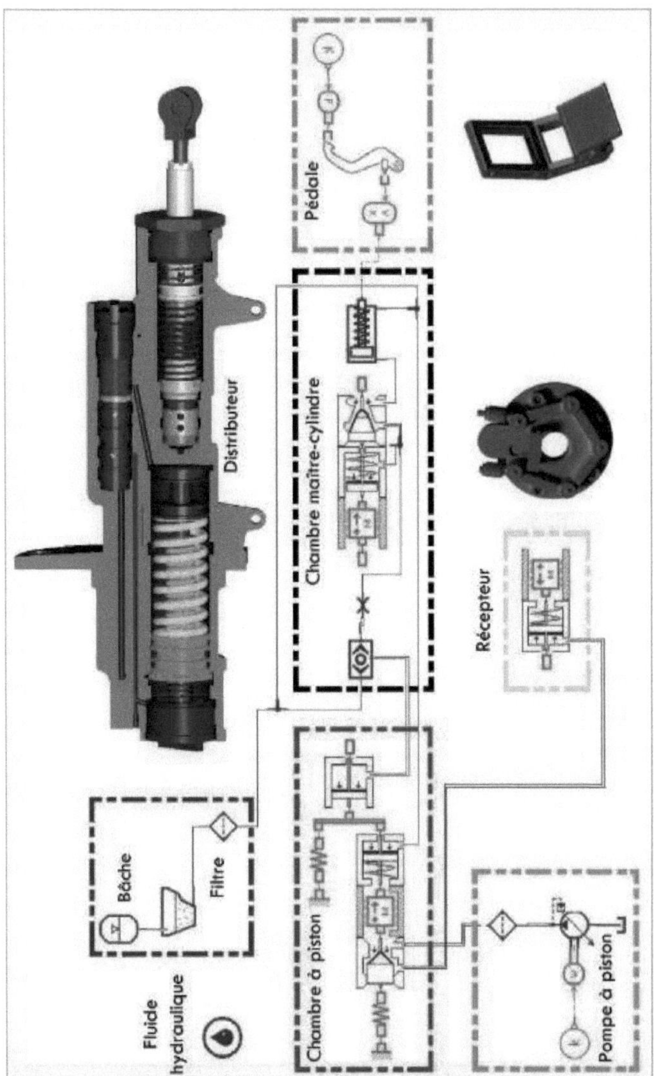

Figure 53. *AMEsim model of the hydromechanical braking system*

To complete our braking system, we have previously developed, in addition to the brake distributor model, 1D physical models of the hydraulic reservoir and fluid, the aircraft pedal and brake control, the hydraulic generation system and the brake block. Causality is achieved by connecting the inputs of one model to the output of another, while respecting the operating principle of the system. The AMEsim tool automatically constructs the equations that characterise the system's behaviour. Hence, the interconnection of these models of the sub-systems making up the braking system produces the development of the complete 1D model of the entire hydromechanical aircraft braking system.

A numerical model of the hydromechanical braking system was established in the AMESim environment in order to create an analysis device for system maintenance and optimisation applications (Figure 53).

4.2.2- Simscape model of the system

With the aim of producing a 1D physical model to simulate the mechanical forces produced by the braking system on the aircraft wheel assembly, we are taking advantage of the specific, high-performance mechanical functions of the Simscape environment to develop this model. To do this, we are using the functional and physical study and the 3D geometric model to design and make concrete the 1D physical models of the sub-systems and the complete hydromechanical aircraft braking system.

In practice, we use components from the Simscape and Simulink special functions libraries, defining the geometric and physical parameters of each component. Then, we proceed to the interconnection of the components which are equipped with energy input and output ports (hydraulic pressure, physical effort, electrical energy...) to build the SimscapelD model of the complete hydromechanical braking system to simulate and analyse the system parameters. We begin by illustrating a common example.

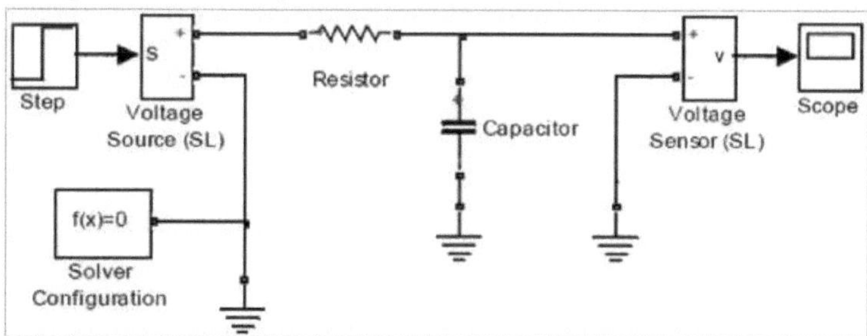

Figure 54. *Simscape model of an RC circuit with generator and sensor*

In this way, the use of *flux* and *effort* variables allows any physical system to be reduced to an equivalent electrical circuit. An effort variable can be lost as a potential and a *flux* variable as a flow rate.

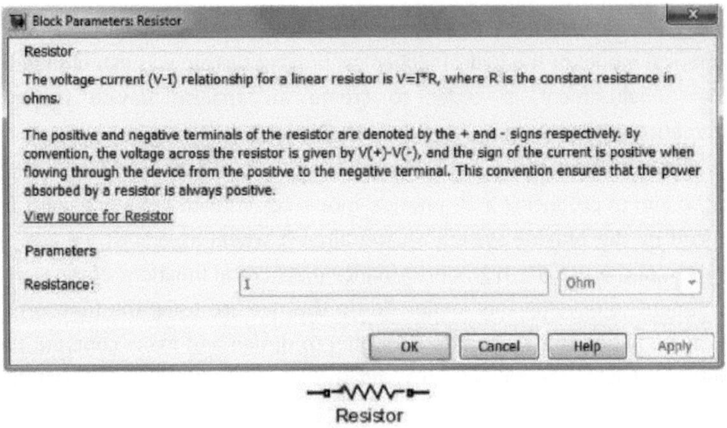

Block Parameters: Resistor

Resistor

The voltage-current (V-I) relationship for a linear resistor is V=I*R, where R is the constant resistance in ohms.

The positive and negative terminals of the resistor are denoted by the + and - signs respectively. By convention, the voltage across the resistor is given by V(+)-V(-), and the sign of the current is positive when flowing through the device from the positive to the negative terminal. This convention ensures that the power absorbed by a resistor is always positive.

View source for Resistor

Parameters

Resistance: 1 Ohm

OK Cancel Help Apply

Resistor

Figure 55. *Simscape Resistor block and mask*

Every Simscape block is associated with a text file in .ssc format which contains the code that determines its operation. Let's take the example of the code associated with the *resistor* block in the *foundation / electrical* library (Figure 56).

```
component resistor < foundation.electrical.branch
% Resistor
% The voltage-current (V-I) relationship for a linear resistor is V=I*R,
% where R is the constant resistance in ohms.
%
% The positive and negative terminals of the resistor are denoted by the
% + and - signs respectively. By convention, the voltage across the
% resistor is given by V(+)-V(-), and the sign of the current is
positive
% when flowing through the device from the positive to the negative
% terminal. This convention ensures that the power absorbed by a
resistor
% is always positive.

% 2011 The MathWorks, Inc.

  parameters
    R = { 1, 'Ohm' };    % Resistance
  end

  function setup
    if R <= 0
        pm_error('simscape:GreaterThanZero','Resistance')
    end
  end

  equations
    v == R*i;
  end
end
```

Figure 56. *Source code for the resistor block*

In Simscape, a *flow* variable is a *through* variable and an *effort* variable is an *across* variable.

Simulating the model in the same Simscape platform allows us to observe and analyse the different physical parameters of the system being modelled. Finally, with Simscape, we can design control systems to visualise the interactions of different physical components in a single system. Thus, modelling in Simscape, as a physical network, has the main advantage of describing the multiphysical structure of a multidisciplinary system rather than the underlying mathematical aspects (Figure 57).

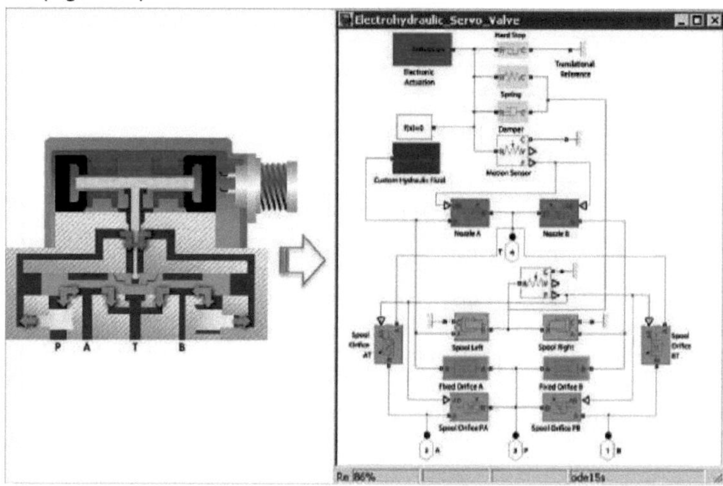

Figure 57. *Example of multiphysics modelling in Simscape*

1D physical model of the aircraft pedal in Simscape

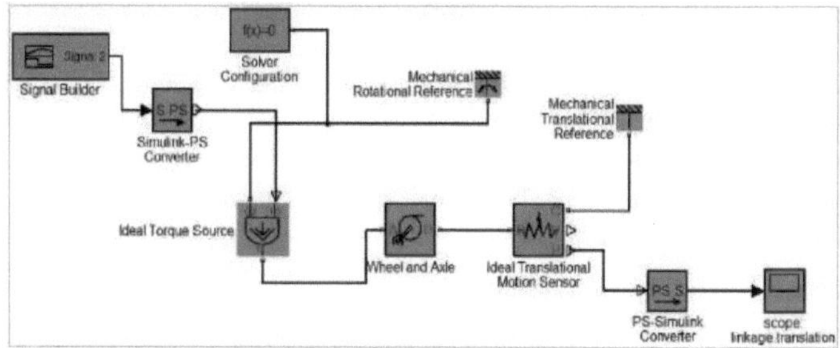

Figure 58. *Simscape model of the aircraft pedal* *.

Legend for Figure 58 :

- Signal Builder: input signal generator for the pedal brake controller ;
- PS-Simulink Converter: converts a physical signal into a Simulink output signal;
- Solver Configuration: specify environment information for simulation

Simscape 1D physical model of the brake distributor

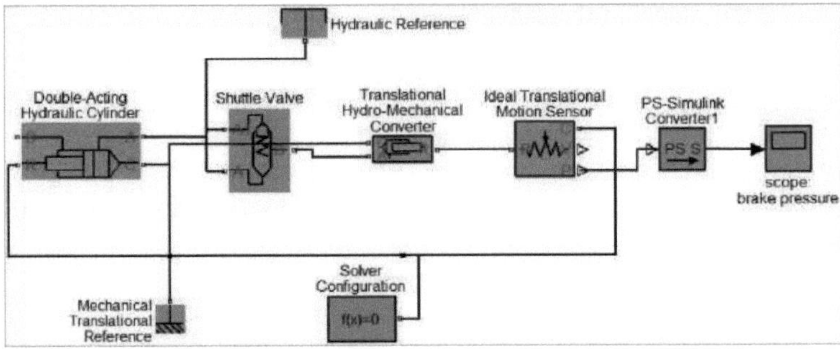

Figure 59. *Simscape model of the brake distributor*.*

and provides parameters for the model's solver;

- ❖ Ideal Torque Source: represents the force acting on the pedal;
- ❖ Wheel and Axle: mechanical link between rotation (pedal) and rotation (axle).

translational movement (rod) ;

- ❖ Ideal translational motion sensor ;
- ❖ Mechanical Reference: blocks that represent a reference point for all mechanical translation and rotation ports;
- ❖ Scope: simulation signal display screen.

Legend for Figure 59 :

- ❖ Double-acting hydraulic cylinder: block representing the hydromechanical translation model, converter, piston chamber;
- ❖ Shuttle valve: block representing a hydraulic shuttle valve;
- ❖ Translational Hydro-Mechanical Converter: an ideal sensor block which converts hydraulic energy into mechanical energy in the form of a translational movement of the converter's output element;
- ❖ Ideal Mechanical Translational Motion Sensor: block representing a device that converts a variable into a control signal proportional to speed and position;
- ❖ Hydraulic reference: block representing a connection to the atmosphere. It has a hydraulic conservation port;
- ❖ Scope: braking pressure applied to the aircraft wheel brake assembly.

Simscape 1D physical model of the aircraft hydromechanical braking system

Modelling the braking system requires the interconnection of the systems: aircraft pedal, brake distributor and wheel assembly (brake block and tyre). We then obtain the model of the hydromechanical aircraft braking system (Figure 60).

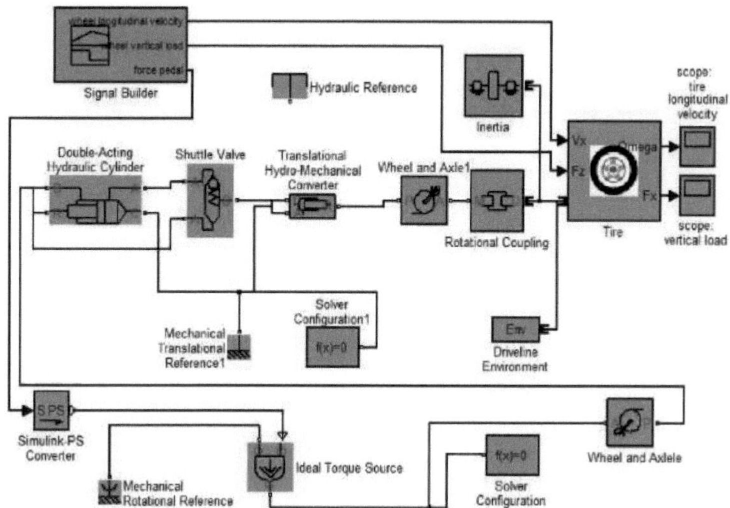

Figure 60: *Simscape hydromechanical aircraft braking model*

Legend for Figure 60 :

- Shoot: block of a wheel with specific vertical nominal load characteristics. Input signals Vx and Fz specify longitudinal wheel speed (m/s) and vertical load (N). The output signals Omega and Fx specify the angular wheel speed (rad/s) and the longitudinal force (n) ;
- Driveline Environment: provides a simulation environment for a block transmission diagram;
- Wheel and axle and rotational coupling: transforms the hydraulic pressure generated by the brake valve into torque applied to the wheel;
- Scope: block which displays the speed of the longitudinal tyre and the braking force.

5- Modelling a multi-domain aeronautical system

5.1- Electrohydraulic braking subsystems

In order to extend the application of this multiphysics approach to component-oriented modelling by adding the electronic component, we return to the modelling of the aircraft electrohydraulic braking system [67]. Using the functional and physical study already described, we proceeded to develop 1D physical models of the excitation, anti-skid and power circuit sub-systems, and thus of the complete aircraft electrohydraulic braking system.

1D physical model of the excitation circuit under AMEsim

The following diagram shows the 1D physical model of the excitation circuit of the electrohydraulic aircraft braking system in AMESim (Figure 61).

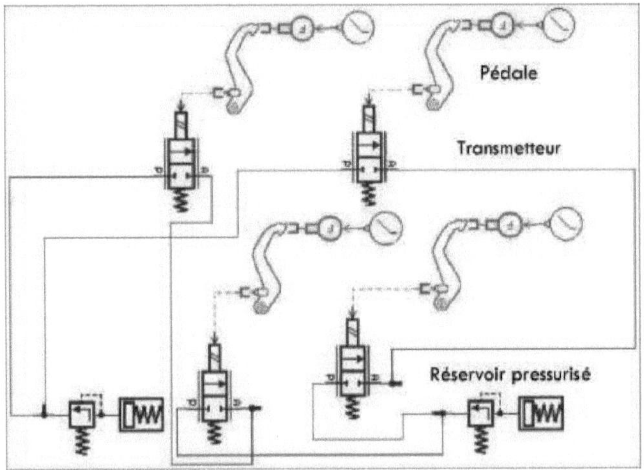

Figure 61. *AMESim model of the excitation circuit*

The excitation circuit shown in the figure comprises :

the four pedals, two in each pilot cabin; the input signals representing the pilot's efforts on the pedals;

mechanical force converters ;

transmitters connected to the pedals which deliver a pressure proportional to the pilot effort;

two pressurised tanks with a spring and two pressure relief valves.

1D physical model of the traction control circuit under AMEsim

In the traction control system, the control unit is the keystone of the chain which receives information from the speed sensors and provides instructions to the solenoid valves associated with each brake block (Figure 62) and (Figure 63).

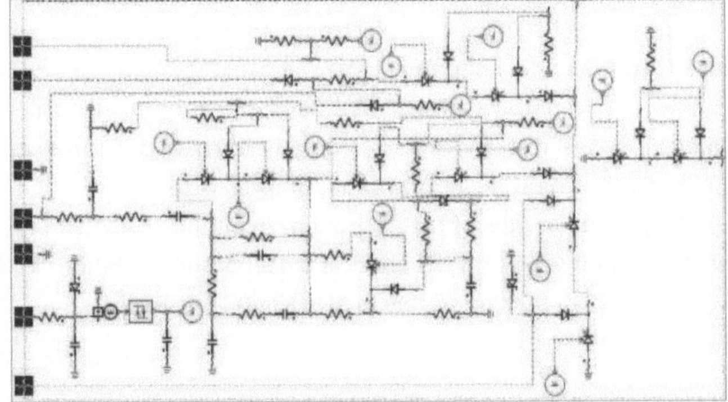
Figure 62. *AMESim model of traction control unit*

The control unit controls the braking of both wheels. The control circuit comprises

for each wheel :
a filter generator modelled on electronic components ;
differentiator (library signal) ;
a trigger represented by the assembly of amplifiers operational ;
a variable discharge memory modelled on a capacitor.
an output stage modelled by the junction of Darlington ;
power supply modelled on a +15V stabilised voltage.

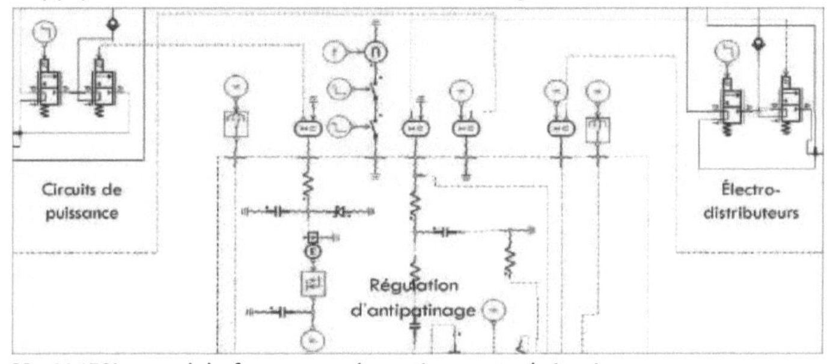

Figure 63. *AMESim model of power and traction control circuits*

5.2- Electro-hydraulic braking system

The multiphysics modelling of the braking subsystems required the connection of hydraulic, mechanical, electrical and electronic components. To develop the 1D physical model of the entire braking system, we interconnect the models of these sub-systems: excitation circuit, power circuit, anti-skid system with control unit, emergency circuit and brake block. We then construct the model of the aircraft's electrohydraulic braking system (Figure 64):

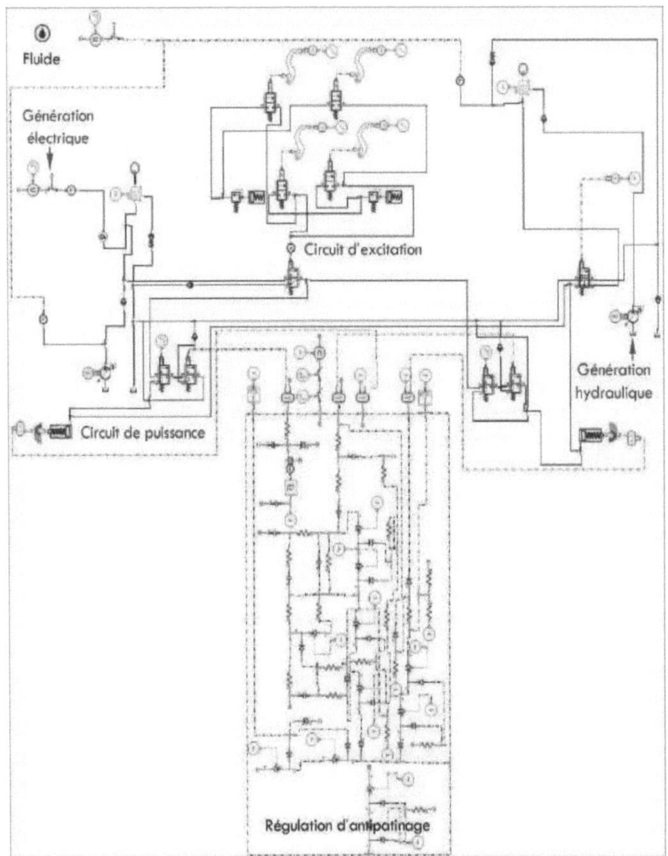

Figure 64. *AMESim model of the aircraft electrohydraulic braking system* 68]

6- Conclusion

A review of the state of the art was presented, focusing on approaches and tools for the design and modelling of multiphysics systems. An evaluation of these approaches has enabled us to identify the strengths and limitations of each of them. In this chapter, a multivalent approach to 1D physical modelling of multi-domain systems has been proposed. This approach is designed using a truly modular process, adopting component-oriented modelling whose tools are compatible with the object-oriented language.

We presented two main tools dedicated to this approach for the development of 1D physical models: AMEsim and Simscape / simulink. This multiphysics approach was applied to the two multidisciplinary aircraft braking systems by developing 1D physical models of their subsystems. This has enabled us to partially address our problem relating to the development of 1D physical models of multiphysics

systems.
On the one hand, the exploitation of the second simulation function of the tools used will enable us to simulate the 1D physical models developed in order to validate these models and to compare their behaviour with the operating principle of the systems in question. On the other hand, in order to improve the study process of aeronautical multiphysical systems, we will suggest a cooperative geometrical-physical co-simulation which will be used to analyse the operating parameters of the systems with a view to optimising the performance of the system's response efficiency.

CHAPTER 5

PHYSICAL - GEOMETRIC CO-SIMULATION - ANALYSIS & OPTIMISATION -

1- Introduction

The design process for a multiphysics system, particularly in aeronautics, must include an analysis of the behaviour of the various parameters of the components, sub-systems and the system as a whole. This analysis is essential because aircraft systems, which are generally considered critical, are designed with reliability and operational safety in mind throughout their life cycle. In order to clarify the notions of safety and reliability of a system, requirements are associated with it and characterise its reliability. In addition, the failure rates of the components making up a system generally do not guarantee the system's overall failure rate. Consequently, the system component redundancy mechanism is often used to reduce its overall failure rate. This mechanism involves using several identical or back-up hardware components to support the same function. Finally, these functional requirements characterised by a probability of failure must take into account the failure rates associated with each of the components and the relative impact of failures on the other related components.

With this in mind, we plan to identify the behaviour of the various physical components and their interactions by launching multiphysics simulations of the 1D physical models of the systems to be studied in the same modelling environments. This simulation will enable us, first of all, to validate the models developed for the systems and their components through the analysis of the functional physical parameters of the correct operation test imposed by the aircraft manufacturer.

Taking into account the limitations of modelling and simulation processes for multiphysical systems, we propose a "cooperative simulation" physical-geometrical co-simulation. In practical terms, after developing the 3D model of the multiphysics aircraft braking subsystem, which is the object of this approach, we will create an interface block connecting the two geometric and physical models. Simultaneous simulation aims to improve the process of studying multiphysics aeronautical systems by taking advantage of the benefits of both physical and geometric simulation processes. In this context, the results of numerical simulations of these models will be used to optimise the behaviour of the system.

2- Towards multidisciplinary 3D simulation

Our approach has its origins in the modelling-based design process, which consists of massively integrating numerical simulation into the development process.

Using simulation models, we can automate part of the design studies, carry out functionality tests and verify and analyse the behaviour of a system in the form of a virtual prototype before going on to manufacture a physical prototype. As a result, the iteration loops of the design process, which we have already described, can arise very early in the development process. In this way, the system reaches a much more advanced stage of development when it comes to physical implementation. In short, an approach based on modelling and simulation has great potential for reducing development costs and times and improving design quality. For this reason, modelling is now commonly recognised as an integral part of the development process [69].

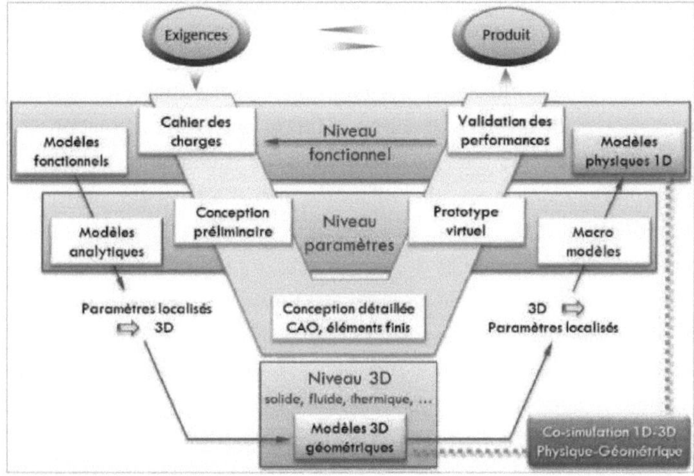

Figure 65. *1D and 3D models in the V-design cycle*

During the development process, different types of simulation model are used (Figure 65). In the top-down phase of the V-process, the system is progressively defined and the level of detail of the models increases accordingly. During this first phase, modelling and simulation are generally used to verify the correctness of the design choices. Initially, we have a functional description of the system to be developed and therefore functional simulation models (software tools: Raphsody, SysML Toolkit, etc.). In the second stage, solution concepts are identified and the corresponding physical or behavioural simulation models are developed. These analytical models can be static or temporal. During the specific design phase, the various components of the solution concepts are defined in greater detail. It is then possible to produce 3D geometric models of these components (software tools: CATIA, Abaqus, Flux3D, etc.). These 3D models can be used to perform local calculations (finite elements) and thus verify the properties of the components developed. Finally, during the system integration phase, the various components are assembled to form the solution concepts and validate their performance. To do this, the highly detailed 3D simulation models are transformed into macro-models with localised parameters, while the 1D physical models of the sub-systems and systems require less detailed simulation procedures.

Usually, for the same level of detail, the different forms of model are linked to different disciplinary fields such as: mechanics, hydraulics, electricity, thermics, etc. This raises the problem of integrating the different types of model into a single simulation environment, in order to take account of the interactions between the different technical disciplines. This raises the problem of integrating the different types of model into a single simulation environment, so as to take account of the interactions between the different technical disciplines. Hence the need to use interfaces that enable a multi-disciplinary approach to design and simulation. However, the lack of communication between abstraction levels, as highlighted above, means that the design cycle for multiphysics products is very onerous and takes longer. In fact, validating and analysing the performance of physical models at the functional level of abstraction independently of the simulation of geometric models developed at the geometric level of abstraction generates considerable losses of material and time. This method is reduced to the analysis of certain parts and aspects of the system and is not adapted to the analysis of multi-domain dynamic behaviour. As a result, it encourages an increase in the number of corrective iterations following a fault detected in the integration tests of the multiphysics component, subsystem or system.

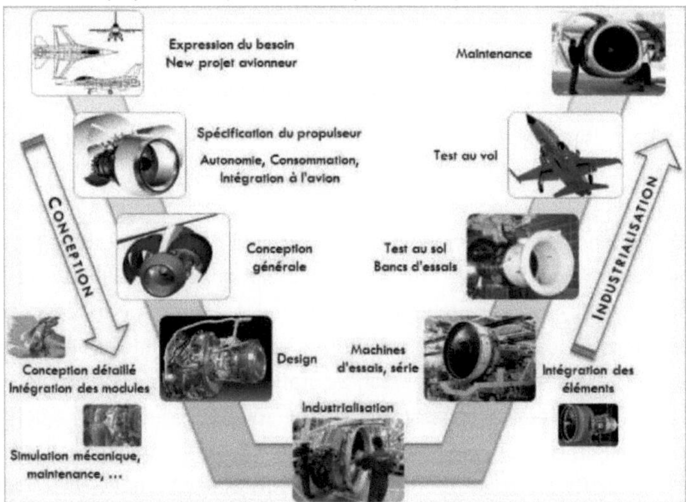

Figure 66. *V cycle: reactor design and production*

Hence the usefulness of a process of co-simulation between the 1D physical models and the 3D geometric models in order to provide a more holistic analysis that takes account of the different physical phenomena in the behaviour of the constituent parts of multiphysics products (Figures 65 - 66). This co-simulation will make it possible to simultaneously run the simulation of the physical model and that of the geometric model. In this way, we can extract the reaction of the mechanical parameters of the parts in the geometric model at the same time as simulating the multiphysics operation of a multidisciplinary component, sub-system or system.

3- Validation of 1D physical models

In order to validate the 1D physical models developed, their simulation and the analysis of the results will be the subject of this section.

3.1- Performance validation process

The assembly of models of the various components constitutes the virtual prototype of the system under study. In the first iterations, a reservation must be made regarding the validation of the models used.

Although the use of valid basic elements significantly reduces the risk of errors, there is nothing to ensure that phenomena that are consciously neglected or ignored by the designers will not disrupt the design stages, thus leading to a repetition of the design cycle. It is the multiplicity of experiences, carefully capitalised on with the help of simulations, that ensures the convergence of the design.

It is therefore essential to validate the models developed before using them for system integration and optimisation.

Real-time simulation is defined by the fact that data exchanges (Input/Output) between the simulator and its environment, illustrated in (Figure 67), must take place at precise instants clocked by a period of time.

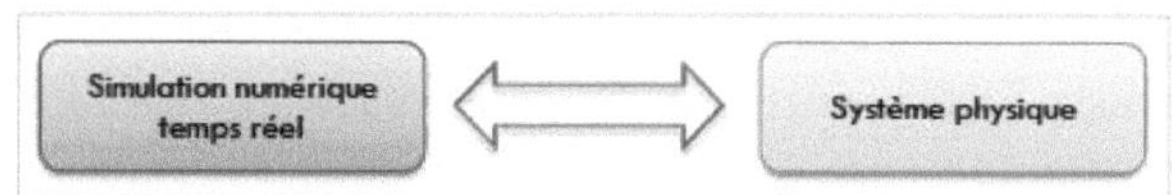

Figure 67. *Hardware in the loop" real-time simulation*

The component-oriented modelling approach based on the AMESim environment makes it possible to simulate in real time the behaviour of variables in the various components of the systems under study.

3.2- Validation of the AMESim hydromechanical model

3.2.1- - Simulation and analysis of functional parameters

The ability to simulate the various physical parameters of the components making up the brake valve allows us, after defining the simulation parameters, to visualise the curves of evolution of their variables over time. This feature allows us to validate the AMESim model of the hydromechanical aircraft braking system (Figure 53) by simulating the behaviour of fundamental system variables such as braking pressure, generated braking friction and exchange rate. We then simulate the test conditions described by the aircraft manufacturer, which relate the input value (pedal force) to the corresponding output value (valve output pressure) [70] :

Hydraulic braking pressure: From Figure 68, we can see that when a force of the order of 70N is applied to the pedal, the valve takes a response time of the order of 0.2s to generate the necessary hydraulic braking pressure.

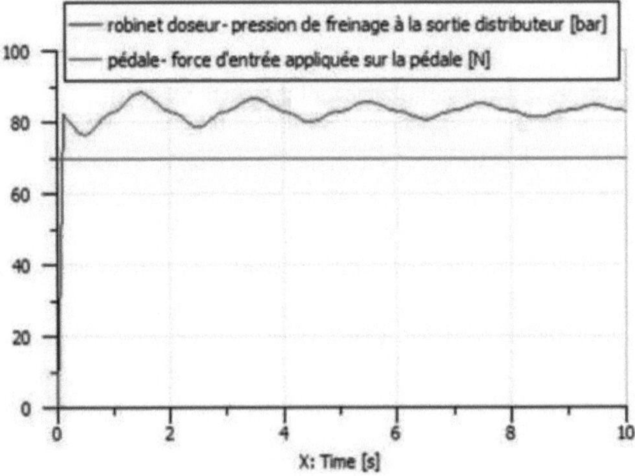

Figure 68. *Evolution of valve output pressure at a braking command of 70N*

The value of this pressure oscillates, with a period of 2s and a progressively decreasing amplitude, around the value of 83bars which corresponds to the generation pressure. This is consistent with the operating principle [55]. In fact, on the one hand, with such a considerable and constant force (pedal fully depressed) leading to a displacement of the valve stem, the pressure generated by the said valve rapidly reaches its maximum 83bars (= the total generation pressure). On the other hand, the oscillation around this value with constant continuity of the same force applied can be explained by the shuttle of the brake block receiving pistons. In this respect, the displacement of the pistons to produce friction between the discs generates an additional volume from which the hydraulic fluid escapes, resulting in a slight reduction in the distributor outlet pressure. The return of the pistons to their initial states following the achievement of the desired braking friction generates an additional return pressure. This results in a slight increase in hydraulic pressure at the valve outlet port.

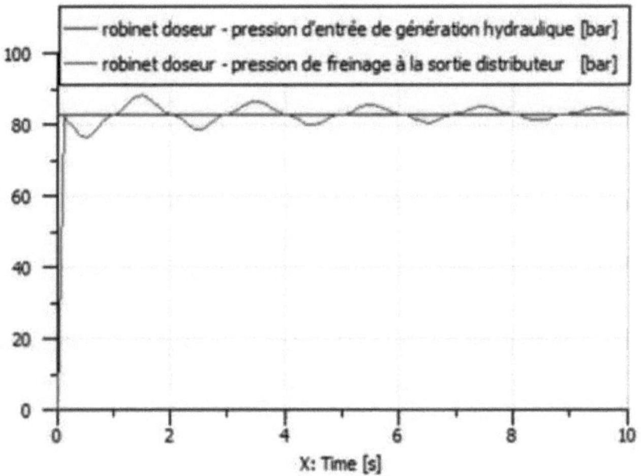

Figure 69. *Changes in spool valve outlet pressure and generation pressure*

Finally, the reduction in the amplitude of this oscillation and the convergence of the braking pressure at the valve outlet towards the value of the hydraulic generation pressure is explained by the tendency of the brake block pistons to assume a position of equilibrium between their chambers which progressively minimises the displacement of these pistons (Figure 69).

Braking friction: The braking force is simply the friction generated at the output of the brake block piston on the brake discs. Its evolution over time with the same value of force applied to the pedal (70N) is illustrated in (Figure 70).

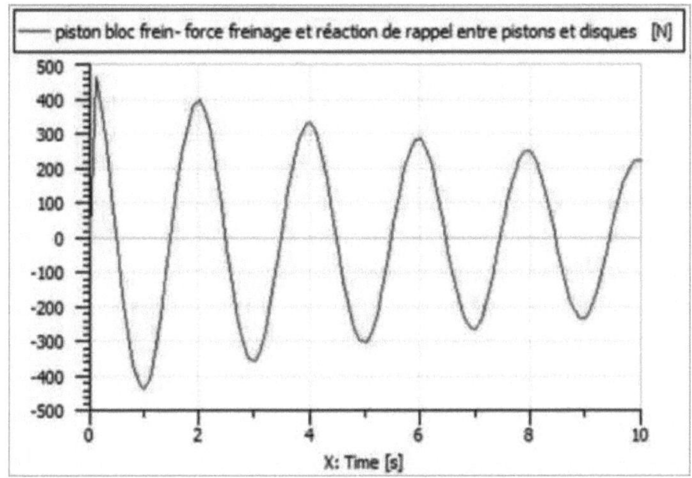

Figure 70. *Variation of the force generated at the output of the brake block piston*

It reaches a maximum peak of around 450N at 0.2s, after which it gradually decreases between positive and negative amplitudes. This evolution corresponds to the action-reaction between the brake block piston and the brake discs via the return spring.

Functionally, the frequency and phase of the evolution of the force are in absolute harmony with those of the evolution of the braking pressure. The analysis of the operation of the subsystems and the results obtained above support the validation of the corresponding models developed.

Hydraulic flow: The metering valve of the brake valve, by construction, maintains a balance of variation in the amount of flow received and discharged. Figure 71 illustrates this theory by observing the phase opposition (180° phase shift) between the inlet and outlet flow of the metering valve with a period of 2s equal to that of the pressure and braking force.

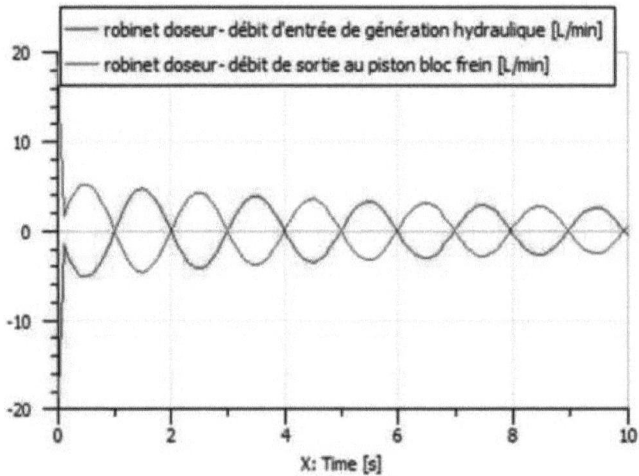

Figure 71. *Changes in inlet and outlet flow rates of the dispenser metering valve*

3.2.2- *Simulation of the aircraft manufacturer's test*

Test assumptions: The purpose of the spool valve product performance test is to determine how the spool valve outlet pressure (braking pressure) changes as a function of spindle displacement (braking control force). This curve, extracted from the manufacturer's system performance manual, is shown in Figure 72. This test recommends the following assumptions [55]:

normal operation (pressure sensing chamber not used) ;
gradual linear increase in braking control ;
constant hydraulic generation pressure = 83bars ;
use of aeronautical hydraulic fluid ;
standard ambient conditions.

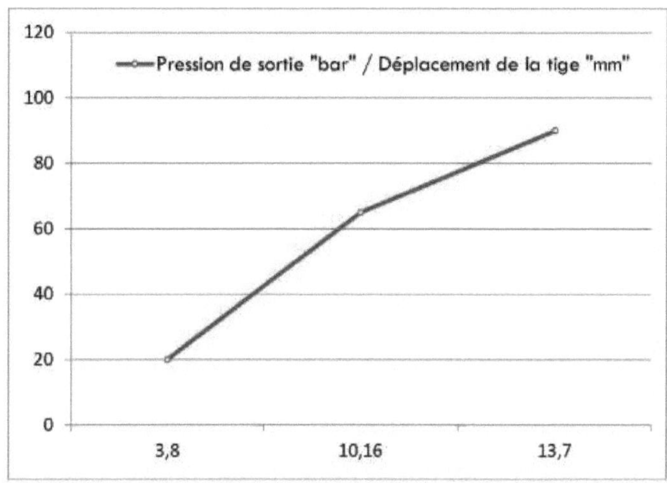

Figure 72. *Essai fournisseur de Pression de sortie distributeur / déplacement de la tige*

Simulation of the experiments on the 1D physical model: We adjust the simulation parameters in line with those recommended by the functionality test and taking into account the hypothesis data on our 1D physical model. We therefore carried out three simulation experiments with rod displacements corresponding to those of the aircraft manufacturer's test:

Experience 1 : Rod displacement = 3.8 mm (Figure 73)

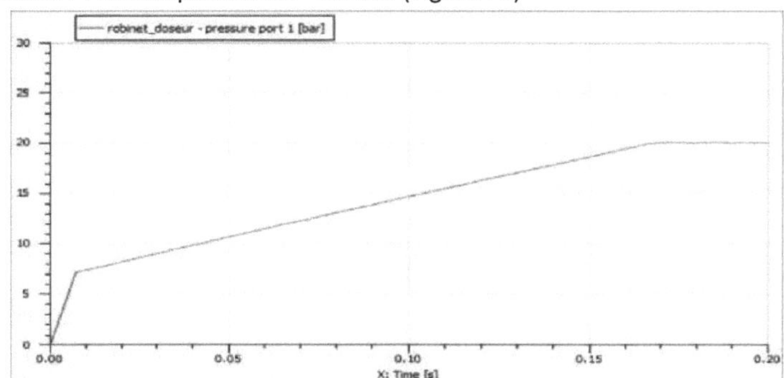

Figure 73. *Spool valve output pressure with spindle displacement = 3.8mm*

Experience 2 : Rod displacement = 10.16 mm (Figure 74)

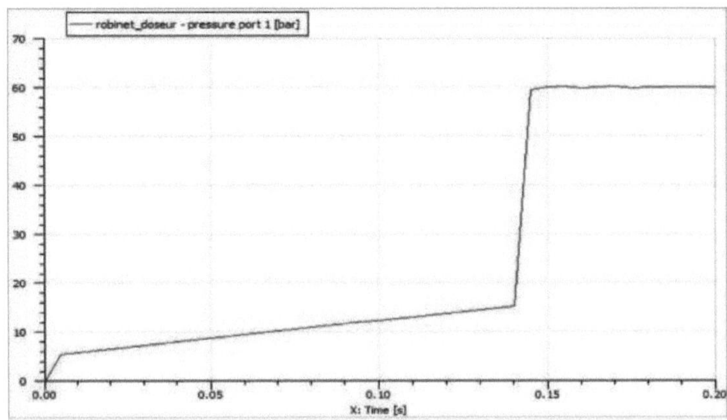

Figure 74. *Pression de sortie du distributeur avec déplacement de la tige = 10.16mm*

Experience 3 : Rod displacement = 13.7 mm (Figure 75)

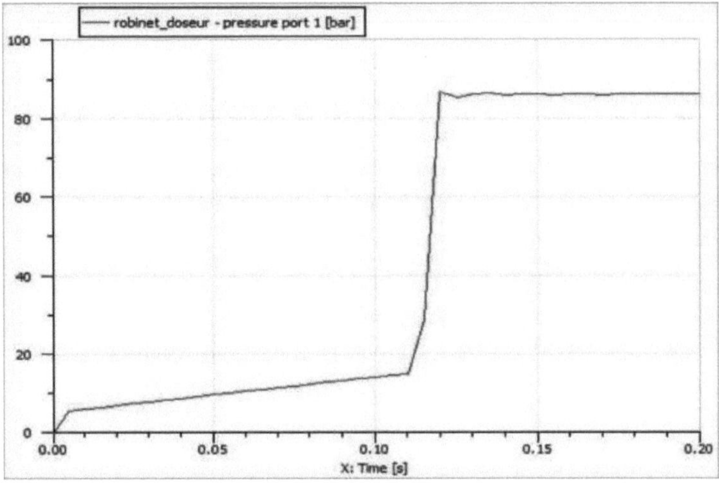

Figure 75. *Spool valve outlet pressure with spindle displacement = 13.7 mm*

From the above, we obtain the resulting curve of brake pressure versus brake command on the 1D physical model. It is illustrated as follows (Figure 76):

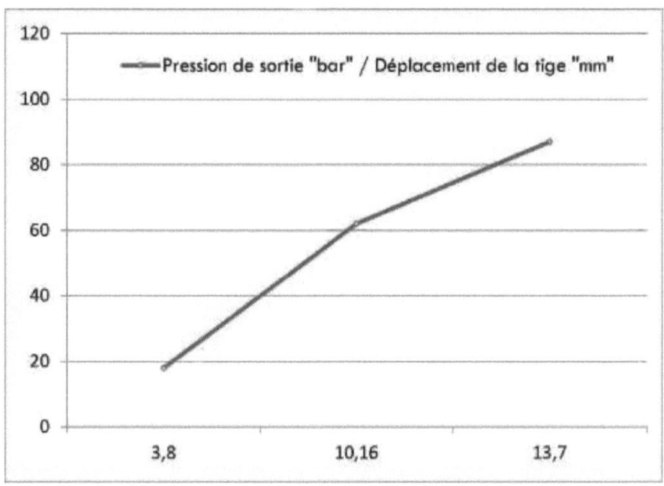

Figure 76. *Evolution of brake pressure as a function of brake command*

Analysis of results: We are putting together the results of the test and the results of the simulation of our model of the evolution of the braking pressure developed by the distributor at the metering valve as a function of stem displacement (Table 2).

	Simulation results			Supplier results		
	Pressure displacement (bars)	Stem (mm)	Output pressure (bars)	Input pressure (bars)	Stem displacement (mm)	Output pressure (bars)
Experience 1	83	3.8	18	83	3.8	20
Experience 2	83	10.16	62	83	10.16	65
Experience 3	83	13.7	87	83	13.7	90

Table 2. *Comparison of results required by the aircraft manufacturer and simulation results*

From Figure 77, we can draw up the comparison charts between the results of the test required by the aircraft manufacturer and our experiments on the physical model:

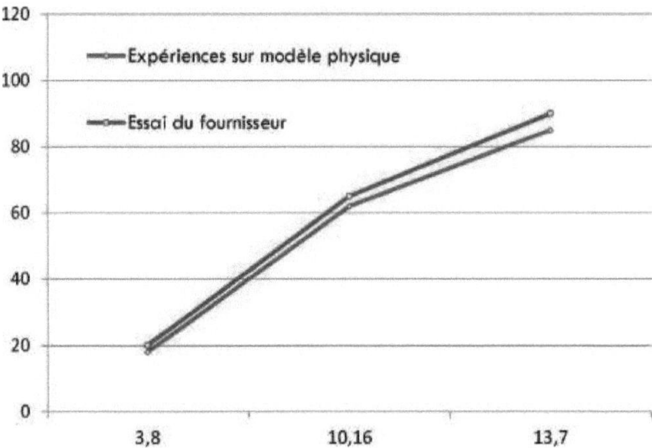

Figure 77. *Abacus for the aircraft manufacturer's test and results of experiments on the 1D model*

There is a slight difference between the results of the supplier's tests and the experimental results obtained on the physical model:

ambient conditions and assumptions taken into account ;

virtual simulation of spring stiffness ;

simplifications on some of the distributor's components;

pressure losses in the pipework and within the distributor, which we have neglected;

high pulsations at the discharge, which increase pressure: need for anti-spill devices;

the aircraft manufacturer's test environment.

It should be emphasised that the functional analysis of the simulation results for key parameters such as braking pressure, braking friction and hydraulic flow reveal that the models developed meet the system's operating principle criteria. Furthermore, by comparing the performance requirements of the test with the results of the corresponding simulations of the 1D physical model, we were able to confirm that our physical models are valid and provide a genuine virtual prototype for analysing the system.

3.2.3- *Simulation and analysis of braking forces*

We simulate the Simscape model of the hydromechanical aircraft braking system, taking advantage of the tool's performance in the mechanical domain to visualise the forces produced by the system on the aircraft wheel. The tool's simulation library allows interchangeable groups of linear and non-linear signal sources to be created and used in the model simulation. In this respect, the signal *builder* block can produce one of several signal shapes: sinusoidal, square, sawtooth and random.

Assumptions and input values: For our model, we choose input signals corresponding to the input parameters of the system. We inject two signals with constant values for the parameters of vertical load and longitudinal speed of the aircraft wheel, and a linear signal ($y = a\ x$) which represents the progressive application of force on the pedal.

Analysis of the results: The parameters to be observed are those related to the braking effect exerted on the aircraft wheel, i.e. the braking force exerted on the wheel. The other blocks are simulated using the default parameter values. The simulation results are shown in Figure 78.

We observe the action of the brake block on the wheel and also the reaction of the return spring of the retaining plate, just after friction is produced, on the discs of the brake block. This action-reaction diminishes over time to stabilise at the initial state of equilibrium before the cycle is repeated. The reason for this braking spacing is to avoid over-braking and heating of the brake block, which could cause the tyre to burst.

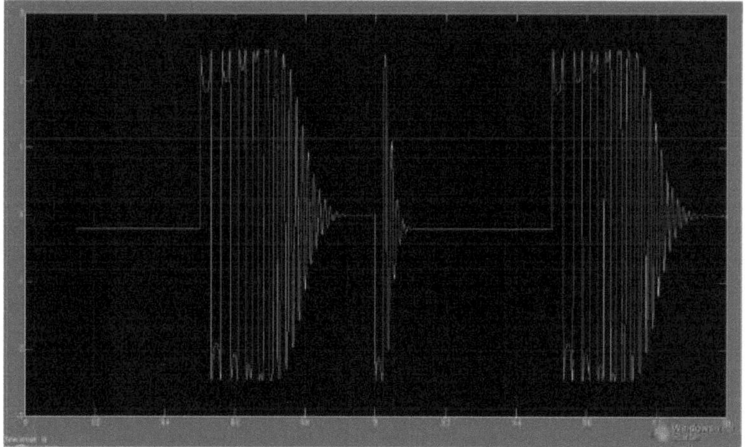

Figure 78. *Output signal for the braking force exerted on the aircraft wheel*

3.3- Validation of 1D electrohydraulic models

3.3.1- Functional recommendations from the aircraft manufacturer

Test assumptions: For the test of this braking system, we assume that the hydraulic generation pressure is assured, which means that the system is operating normally [43].

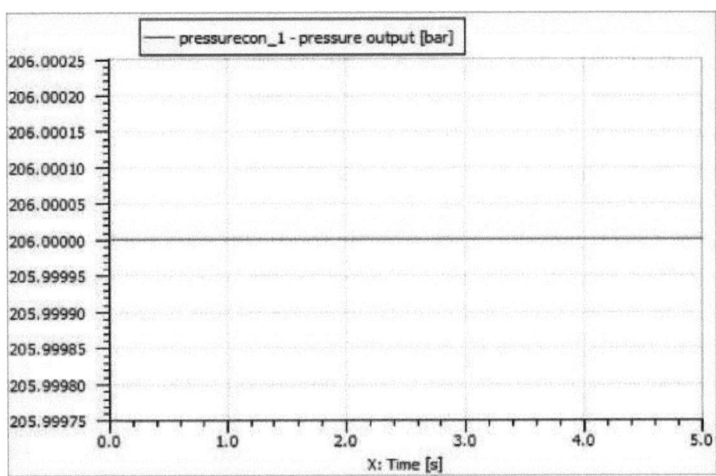

Figure 79. *Hydraulic generation pressure*

We have two input parameters:

Hydraulic generation pressure: this is constant, at around 206±2bars (Figure 79) ;

The force applied to the pedal: this increases progressively along a straight line as a function of time until the "pedal to the floor" position is reached (Figure 80).

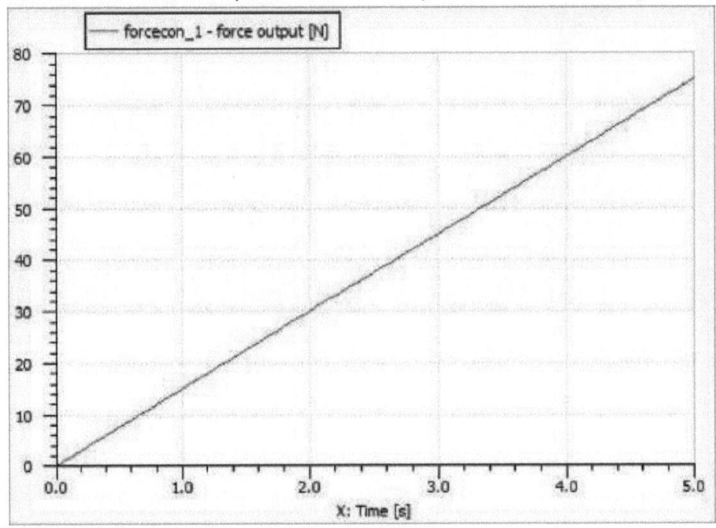

Figure 80. *Evolution of the force applied to the pedal*

Test performance requirements: The hydraulic braking pressure generated by the power circuit to the brake block must be between 100bars and 130bars with the pedal fully depressed; The normal accumulator inflation pressure must be 206bars.

3.3.2- *Validation of the 1D electrohydraulic model*

The results of the simulation of the output parameters observed after the simulation of the 1D physical model of the electrohydraulic braking system are the output pressure of the excitation circuit and the output pressure of the power circuit directed to the wheel brake block assembly.

After the progressive application of the pilot force on the pedals, the pressure of the excitation circuit increases linearly ($P_{excitation} = 30t$) for a period of 0.5s to generate a stable excitation pressure of 15bars regardless of the progression of the force applied to the pedal. This explains why 15bars is the normal operating pressure of the excitation circuit delivered to the power circuit (Figure 81).

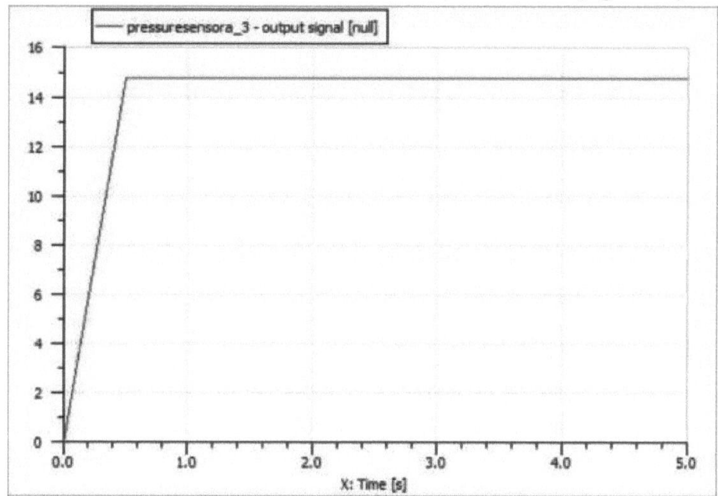

Figure 81. *Excitation circuit pressure*

We observe that the pressure in the power circuit progresses linearly over a period of 0.8s with two different directrix coefficients in two phases. The first phase is before reaching 60bars with a slow evolution (directing coefficient = 80) and the second phase is above this value with a critical evolution (directing coefficient = 500). The pressure then peaks at 122bars before stabilising at 120bars with a response time of one second. When the pedal is fully depressed, the pressure delivered by the braking system is of the order of 120bars. Depending on the value required by the test, the hydraulic braking pressure, measured with the pedal fully depressed, should oscillate between 100 and 130bars.

The braking pressure of our 1D physical model falls within this range (120bars). Thus, the comparison between the results obtained and the performance required by the aircraft manufacturer confirms the validation of the model developed (Figure 82).

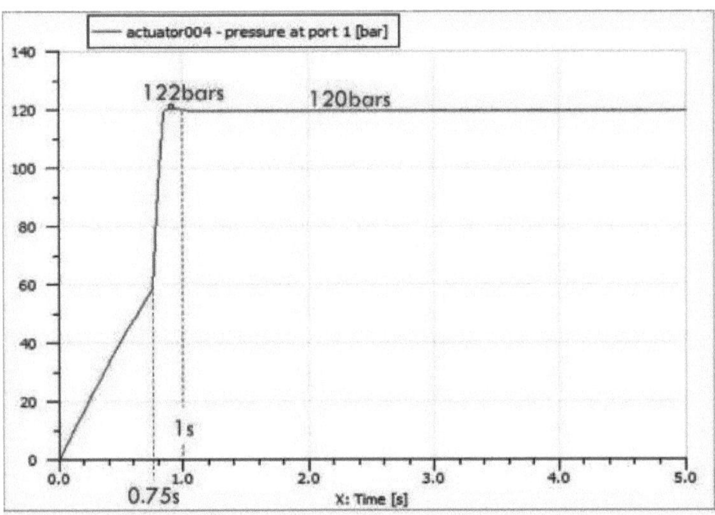

Figure 82. *Pression hydraulique de freinage générée par circuit de puissance*

In this section, we verified the validation of the developed physical models of the electrohydraulic aircraft braking system via the simulation of the test run by adjusting the appropriate simulation parameters of the models in the AMESim multiphysics simulation tool. This simulation enabled us to confirm the validation of the models and also the extraction and analysis of the evolution of the various multiphysics parameters of the system, in particular the braking pressure.

In the next section, we propose a process for developing a co-simulation coupling the physical model and the geometric model of a multiphysics subsystem under an interfacing platform.

4- Co-simulation of a multidisciplinary system

4.1- Interfacing with modelling software

Insofar as there is currently no single software package capable of covering all the areas mentioned and all the levels of design abstraction, it is essential to be able to interface the software packages chosen on the basis of their complementarities. There are two main forms of interfacing:

Code export: This involves exporting the code from software A to software B, with the latter's solver having the role of solving the complete system. Since the software does not use the same model description language, adaptations (interfaces) must be developed from software to software. Here we can see the advantage of having a common language that appears to meet user expectations (trends: VHDL-AMS and MODELICA). However, they are unable to agree on a universal solver that solves this interfacing problem.

Co-simulation: The tools involved in the simulation use their executable files and run their own solvers separately, exchanging data during the simulation. System A is solved using A's solver and system B is solved using B's solver, with information being exchanged between the two systems at predefined time intervals. This co-simulation technique is difficult to use when systems A and B are described by systems of algebraic-differential equations.

4.2- The benefits of physical-geometric co-simulation

The task of developing models for generic multiphysics dynamic systems is usually complex because of the requirement for specialised simulation tools for functional analysis and physical design. However, this work becomes more complex for simulation accessibility and dynamic analysis models or for the control of multiphysics prototypes. Indeed, such models must be able to represent all dynamically relevant effects.

Currently, physical simulations of mechanical systems are not affected by the actual geometry of the objects. Most simulation systems represent mechanical parts by their centres of mass or by a few geometric characteristics such as moments of inertia. In fact, current modelling and simulation tools do not allow the different levels of abstraction to be processed with a view to analysing the dynamic and functional parameters and also their interactions (Table 3).

Levels of abstraction				
Software Publishers	Functional	System	Sub-system	Geometric
Matlab / Stateflow MathWorks	J			
ASCET ETAS	J	J		
Matlab / Simulink MathWorks		J	J	
Simscape MathWorks		J	J	
Dymola Dassault systemes		J	J	
LMS Imagine.Lab AMESim /LMS		J	J	
OpenModelica Modelica Association		J	J	
ProEngineer PTC				J
CATIA / Abaqus Dassault Systemes				J
Fluent ANSYS				J

Table 3. *Examples of software tools according to the level of abstraction treated*

As a result, neither the geometric nor the physical model of the multiphysics system can generate the simultaneous evolution of the physical parameters, taking into account both the physical and geometric parameters and their simulation. As a result, designers always have to separate the geometry from the physical model, depending on the level of abstraction involved. This separation makes the simulation-oriented design of a multiphysics product slow and inconsistent.

One significant example of multiphysics systems suffering from this handicap is hydromechanical systems, such as hydraulic chamber enclosures. These are continuous non-linear dynamic systems characterised by the presence of hydraulic fluid. Clearly, dealing directly with three-dimensional models without taking into account the major effect of fluid pressure is not practical for both dynamic analysis and simulation. Consequently, it is necessary to introduce methods for describing the dynamics of such a system.

In order to overcome this handicap, we propose a co-simulation process that simultaneously links the design abstraction levels. This process consists of developing an interface linking the 3D geometric model to the 1D physical model. This interface will enable us to jointly simulate the various dynamic and multiphysical parameters. Such a process will be applied to a hydromechanical enclosure of the brake distributor where the simulation of the dynamic reaction of the geometric parts to the presence of the hydraulic fluid will present the strong point of this co-simulation.

4.3- 1D - 3D co-simulation process

4.3.1- Virtual.Lab Motion LMS interface

LMS Virtual.Lab's Motion interface is a platform for interfacing the language of its Motion 3D simulation tool with that of 1D physical simulation tools for simultaneous simulation analysis. This interface option between multiphysics modelling software is very useful for studying the interaction between the 3D model of a complex mechanism and sophisticated pneumatic and hydraulic circuits. The interface makes it possible to plot the results of each system in its own package and to take full advantage of the animation capabilities of LMS Virtual.Lab (Appendix 2).

4.3.2- Interfagage Motion- AMESim

We are using the Motion platform interfacing capability of LMS Virtual.Lab to leverage the capabilities of Imagine.Lab AMESim and Virtual.Lab without diminishing the performance of either package. This arrangement will allow us to link the 3D geometric model of a brake valve mechanism built in the Virtual.Lab Motion environment (or CATIA import) to a 1D physical model in AMESim. This interface between Virtual.Lab Motion and AMESim can support the following types of analysis:

Coupled co-simulation: The Motion-AMESim coupled co-simulation process is a master/slave relationship during which all Motion equations and AMESim state equations are combined and solved simultaneously. This also allows the use of the static Motion solver and linearisation. Different types of analysis are supported in the coupled cosimulation process: static, dynamic, static/dynamic and linearisation (Table 4).

static solution: is an algebraic solution to the equations of state Motion such that the state derivatives are zero and the system is at a set point of equilibrium;

dynamic solution: in this solution, Motion's fundamental differential-algebraic equations are solved using a numerical integration process;

static / dynamic solution: in this analysis solution, first a static solution is made to provide the Motion a equations.

A dynamic solution is launched from the equilibrium position;

linearisation: linearisation analysis is part of the dynamic solution. At the point of reporting in the solution process, the system is converted to linear analysis. The linear representation is useful for control design.

Type of interface	Lead solver	Calculation tool	Types of analysis
Coupled co-simulation	Motion	Motion	* Static * Dynamic * Static / Dynamic * Linearisation
Dynamic co-simulation	AMESim	AMESim	* Dynamic

Table 4. *Motion interface analysis types*

In coupled co-simulation the equations of state of both AMESim and Motion are solved as a complete set by the Motion simulation solver. The Motion solver creates a set of Motion algebraic differential equations which are represented as Newton-Euler equations. Virtual.Lab uses a maximum set of coordinates, then removes additional degrees of freedom by applying a library of link constraint equations:

$$M\dot{v} + \Phi_q^T \lambda = Q_a(q, v) \qquad (3)$$

$$\Phi(q) = 0 \qquad (4)$$

Or :

q is the vector of generalized position coordinates ;

v is the vector of generalized coordinate speeds ;

M is the mass matrix ;

Q_a is the vector of applied forces;

λ is the vector of Lagrange multipliers.

When Virtual.Lab Motion and AMESim are coupled, an AMESim force control assembly is applied to the bodies of the mechanism resulting in Motion. With the operation of the load position sensors, velocity and acceleration data are fed to

AMESim. Typically, AMESim forces are the product of equations of state and Motion must integrate a set of AMESim differential equations. The AMESim state equations are represented by "g" and the state variable by "x", so Motion's coupled equations become:

$$M\dot{v} + \Phi_q^T \lambda = Q_a(q, v, x) \quad (5)$$

$$\Phi(q) = 0 \quad (6)$$

$$g(q, v, x, \dot{x}) = 0 \quad (7)$$

The AMESim equations are written in a general format to allow the implicit equations to be generated and solved by the Motion solver. Each model is built with its software solution, then the two models are linked in the Motion workbench and coupled co-simulation is performed within Motion or using the loader package. Figure 83 shows the general interfacing process:

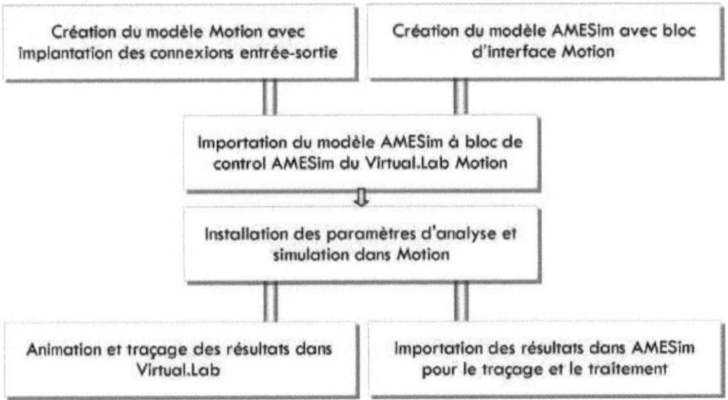

Figure 83. *Coupled co-simulation process between AMESim and Motion*

Dynamic co-simulation: In the dynamic co-simulation approach, in contrast to coupled co-simulation, each simulation package (LMS Imagine.lab and LMS Virtual.Lab) runs its own solver in synchronisation with the other solver, and the equations of motion from AMESim and the equations of state from Virtual.Lab Motion are solved separately.

To do this, in dynamic co-simulation mode each solver switches on and communicates with the other solver at discrete time intervals. The same equations (5) and (7) are solved in dynamic co-simulation mode as in coupled co-simulation, but separately. In this situation the AMESim solver is the master and solves its own set of state equations from instant "t_i" to the instant at the next communication interval "t_{i+1}".

Equation (7) becomes equation (8) where Motion's input variables (q, v) are always at the last time sampling instant.

$$g(q^i, v^i, \chi, \dot{\chi}) = 0 \qquad (8)$$

Once the AMESim solver has completed the integration to the next communication interval the Motion Solver is called to integrate at the current time. The Motion solver now uses the AMESim inputs at the last communication interval "t_i" to advance to the next communication interval "t_{i+1}".

$$M\dot{v} + \Phi_q^T \lambda = Q_a(q, v, \dot{\chi})$$

$$\Phi(q) = 0$$

There is a similarity between coupled co-simulation and dynamic co-simulation. However, in the dynamic co-simulation mode, the Motion model becomes a "sub-component" of AMESim (Figure 84):

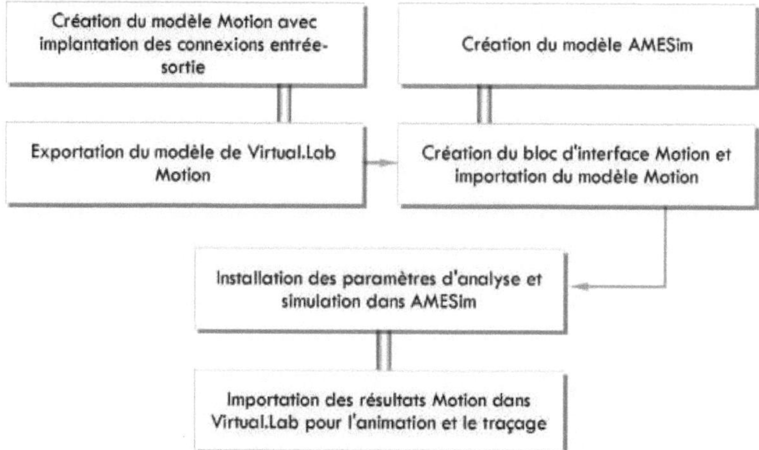

Figure 84. *AMESim-Motion dynamic co-simulation process*

5- Implementation of co-simulation and optimisation

5.1- 1D-3D co-simulation of a multiphysics system

Given that our brake valve has three chambers which communicate via hydraulic fluid, through interconnections which cannot be modelled geometrically, we are interested in the dynamic study resulting from this co-simulation in the master-cylinder chamber of the brake valve. This chamber is an essential hydromechanical mechanism in the operation of the distributor, hence the interest of this dynamic study of the reaction of the constituent parts of this chamber to the displacement of the rod as a braking control element.

We draw inspiration from the coupled co-simulation and dynamic co-simulation processes to carry out physical-geometric co-simulation (1D3D) using the Motion platform from LMS Virtual.Lab and AMESim. This process is also applicable to other system modelling tools, such as Simscape / Simulink, Modelica, etc.

5.1.1- *Coupled co-simulation process*

Development of CAD models of component parts and assembly using the Motion interface (Figure 85) and (Figure 86) :

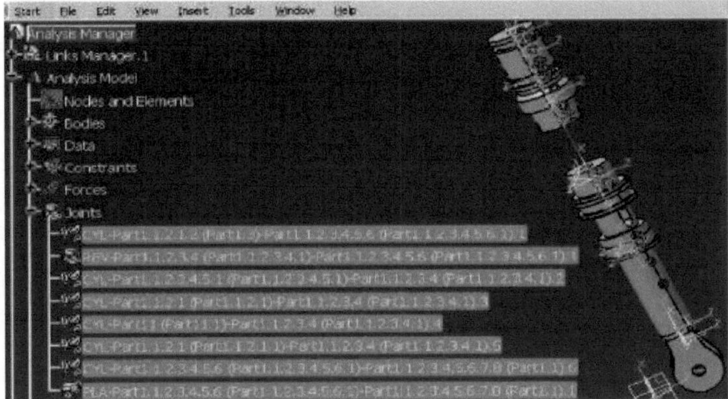

Figure 85. *Definition of connections between parts*

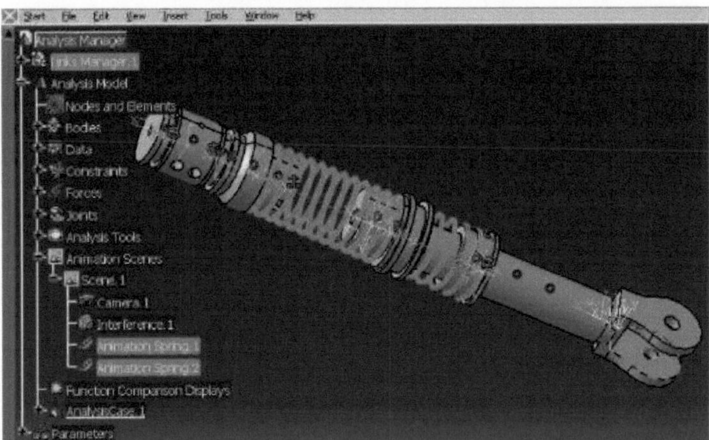

Figure 86. *Inserting parameter springs*

Definition of the points on the parts to be studied as inputs to the coupled simulation (Figure 87). In general, we take a reference point on a fixed body and the second point on the part we wish to study its dynamic and kinematic behaviour with the presence of hydraulic fluid in the chamber and also under the constraint of a rod displacement as a braking control.

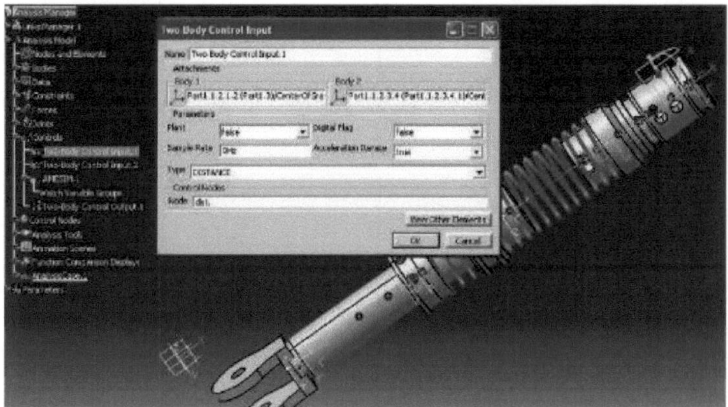

Figure 87. *Definition of inputs*

We insert from the Motion platform the 1D AMESim model of the brake distributor (Figure 88):

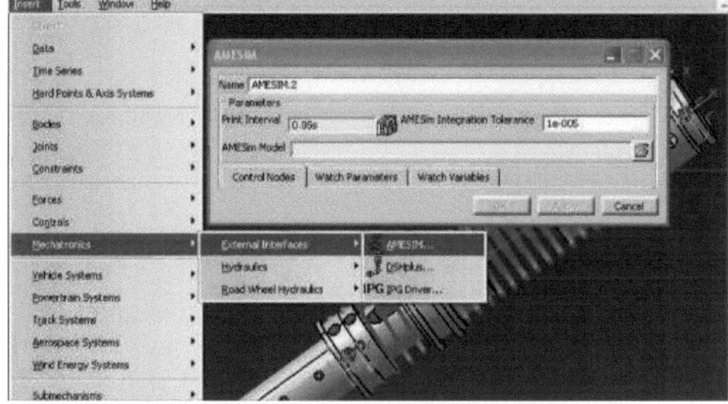

Figure 88. *Inserting the AMESim model*

The link between Motion's 3D model and AMESim's 1D model (Appendix 3), for the planned coupled co-simulation, is established by creating an interface block which links the two models. We specify for this block the input parameters and the output parameters that we wish to simulate (Figure 89):

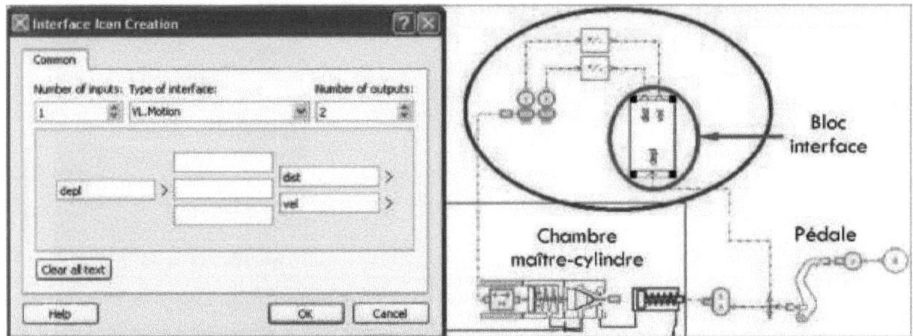

Figure 89. *Creating an interface block*

We define the type of dynamic analysis and the analysis parameters (Figure 90):

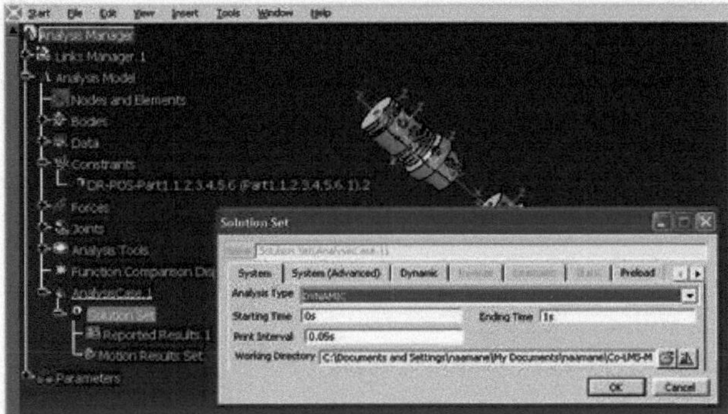
Figure 90. *Type of analysis Dynamic*

We choose the types of interference to be displayed during the dynamic simulation of the system in the presence of hydraulic fluid (Figure 91):

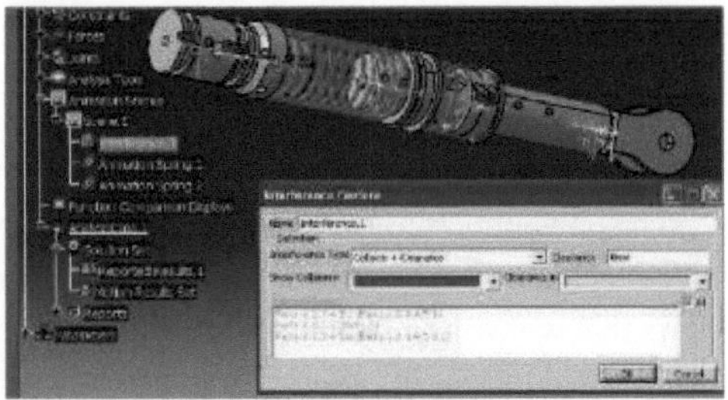
Figure 91. *Definition of the interference to be studied*

5.1.2- Co-simulation and results analysis

We run the co-simulation in the Motion interface. We can visualise the interferences with the corresponding tolerances (Figure 92):

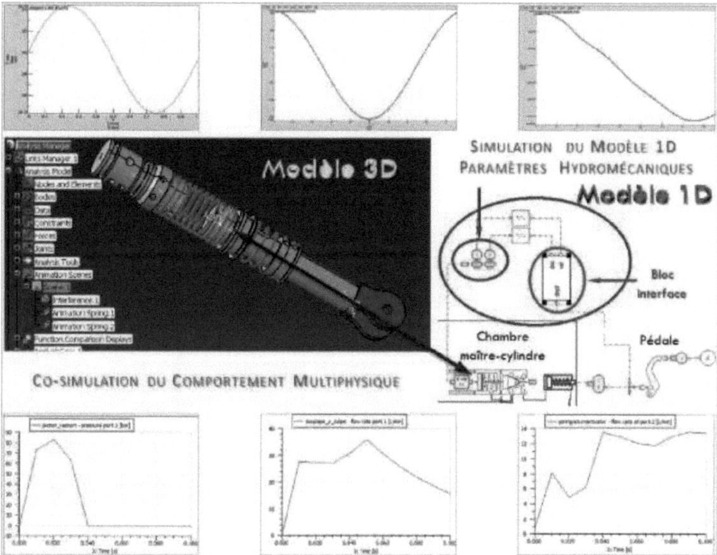

Figure 92. *Interference simulation*

The main benefit of this type of co-simulation is that the dynamic parameters of the parts in the 3D CAD model can be visualised while the system is operating on a multiphysics basis. The results of the co-simulation are extracted by applying the above process.

We run the 1D physical model compilation and analyse the dynamic reaction of the master cylinder parts in the CAD model from the output of the geometry-physics interface block. We visualise the evolution of the displacement of the valve stem following a braking command corresponding to an initial displacement of the order of 3cm (Figure 93):

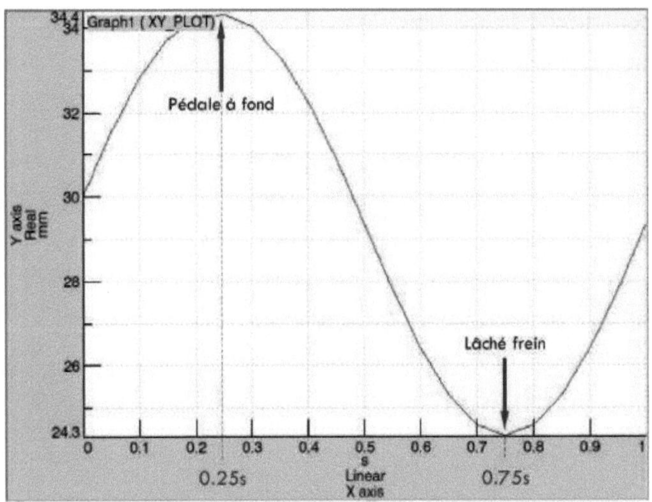

Figure 93. *Déplacement de la tige*

The rod moves in a sinusoidal shape between the maximum position (maximum braking pressure) and the minimum position (brake release) in which the rod is depressed. This sinusoidal shape is explained by the action-reaction of braking pressure and return pressure acting on the rod. We also note that the desired braking is achieved at 0.25s and the rod returns to rest at 0.75s.

We can also see the evolution of the speed of movement of the valve spring and also the torque exerted on this spring (Figure 94):

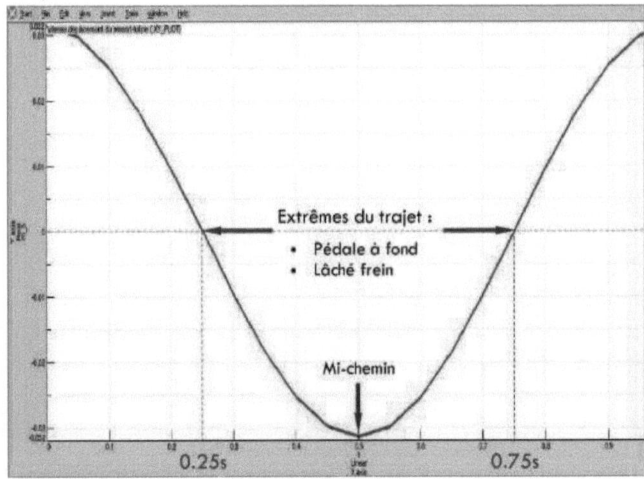

Figure 94. *Displacement speed of the valve spring*

We note that this velocity has a maximum of the order of 0.032 m/s in both the extended and compressed positions. Its value also evolves sinusoidally over a period of one second.

The evolution of the value of the torque exerted on the spring corresponds exactly to the evolution of the displacement of the rod as shown in (Figure 95). The extreme values of the torque exerted correspond to the extreme displacements of the rod.

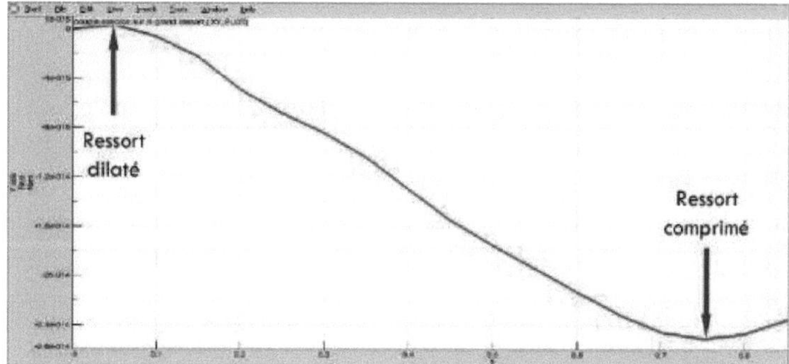

Figure 95. *Torque on the spring*

Figure 96 shows the energy developed by the rod as a result of the torque applied. We can see that the period of devolution of the value of this energy is two (0.5s), so it is cancelled out twice before the rod returns to the released brake position.

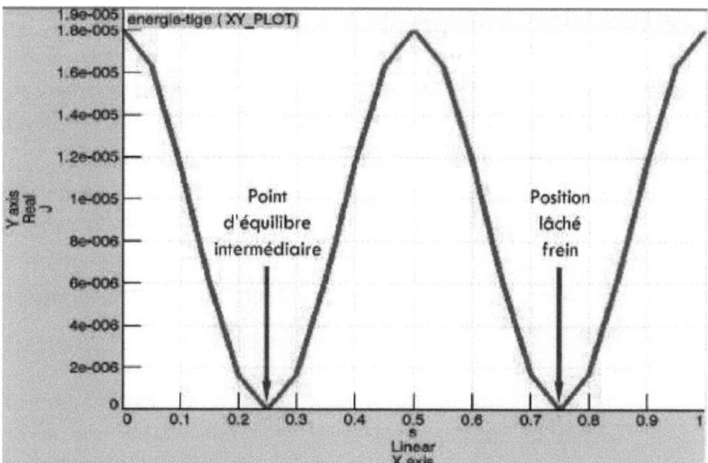

Figure 96. *Energy developed by the stem*

Other values can also be obtained and analysed on the different parts of the 3D geometric model, such as the accelerations of part displacements [71].

Modelling and simulation of multiphysical aircraft braking systems can also help to predict the maintainability of these systems or their components and, above all, to optimise their behaviour for operating efficiency.

5.2- Optimising system response time

The modelling and simulation of the multiphysics aircraft braking system will contribute to the optimisation of the behaviour of such a system. This objective will be the subject of this section.

5.2.1- Optimum design

Optimal design, also known as shape or structure optimisation, is a key concern for engineers and researchers who design innovative industrial products (multiphysics components, mechanical structure, aerodynamic profile, electronic components, etc.) in order to improve essential physical properties (efficiency, strength, durability).

This design must also satisfy strict constraints such as cost, industrial feasibility, but also the weight or volume of the structures, or any other important physical property. It is therefore a classic problem in operational research: the constrained optimisation of an objective function.

Given the computing power of computers, mathematics has become essential in automating this optimisation process. Indeed, the traditional method of optimisation was to proceed by trial and error, following the know-how and intuition of the engineer: we try out a shape whose performance we calculate, then, depending on the latter, we modify it to try to improve it and we repeat the process until we obtain a satisfactory (or even heuristic) shape. This 'manual' approach seems very slow, expensive and imprecise. Increasingly, it is being replaced by numerical optimisation software that represents the shape using a limited number of descriptive parameters (usually control points on the edges), and iteratively improves it by automatically varying these parameters.

This is where the algorithms are hidden, sometimes ancient, but more often extremely 'cutting-edge' and the result of contemporary research. In particular, in the context of mechatronics, especially automotive and aeronautics, recent developments in geometric and topological optimisation of shapes have had a considerable impact on industry.

5.2.2- Optimising response time

As we have already mentioned, the braking pressure delivered by the power circuit only reaches its functional pressure of 120bars after one second of evolution. This time represents the response time of the aircraft braking system and is considerable for a fighter aircraft characterised by its high landing speed, taking into account the constraint of the length of the runway.

In order to reduce this brake pressure response time, we carried out an in-depth study of the functional parameters of each system component and their effects on the parameter in question by means of scenario simulations.

In this perspective of braking efficiency, which results in a reduced and optimal 'response time', the said study and repeated simulations have revealed that the operation of the pressure relief valve installed on the hydraulic reservoir, whose role is to bleed the reservoir pressure, has a negative influence on the rapid increase in excitation pressure.
In order to optimise the braking pressure response time of our physical model, we have studied the parameters of the pressure relief valve and also the capacity of the associated hydraulic reservoir in order to provide an optimisation solution that maintains the correct operation of the system. We acted on the hydraulic pressure thresholds for triggering the pressure relief valve, while compensating for this change by increasing the capacity of the reservoir in the excitation circuit.
We adjust the new parameters on the physical models of the pressure relief valve and the hydraulic reservoir. Simulation of the modified 1D physical model of the braking system has enabled us to visualise our "response time" parameter.
The result is considerable insofar as we can reduce the response time by 50% (0.5 seconds) with a moderate evolution which favours reliable operation (steering coefficient = 250) and above all while respecting the other operating performance requirements (braking pressure = 122bars) (Figure 97).

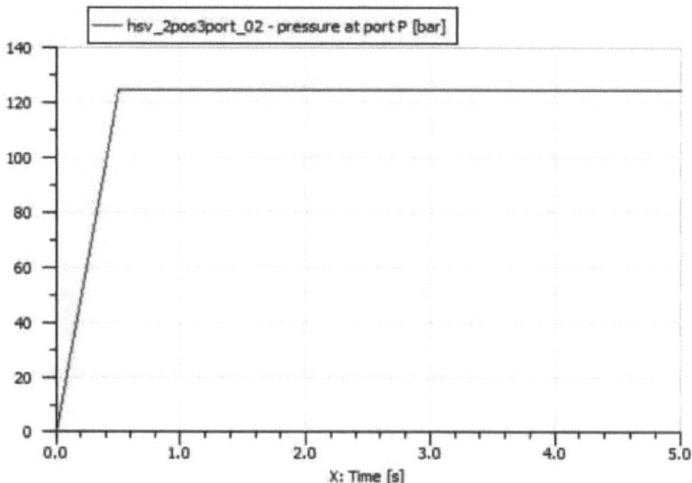

Figure 97. *Optimisation of braking pressure response time*

The simulation of 1D physical models of multiphysics aircraft braking systems in two environments based on the component-oriented approach consists in analysing the variation over time of the dynamic parameters of the subsystems involved in aircraft braking systems. 1D physical models have been used to predict the behavioural characteristics of the systems under different operating conditions. In fact, by simulating the determining parameters of the systems, we were able to validate the models developed by highlighting the operating principle of the system. It also enabled us to analyse the system performance parameters by simulating the test run required by the aircraft manufacturer. Finally, we can visualise the evolution of the force produced by the system on the wheel in the form of an action-reaction between the brake block and the wheel as a parameter describing the dynamic behaviour of the aircraft wheel on the ground under brake stress.

The results of the simulation led us to optimise the parameter determining the response time of the braking system. This optimisation was the basis for an in-depth study using repeated simulations to determine the parameters influencing the system's response time.

6- Conclusion

The simulation of the 1D physical models of the aircraft braking systems enabled us to validate these models by referring to the simulation of the functional test required by the aircraft manufacturer. We were then able to re-examine the operating principle of the braking system. However, the conceptual study of the behaviour of such a multiphysics system will only be optimal if we are able to carry out the different types of functional tests in the 1D virtual environment in harmony with the 3D geometric virtual model. In response to this problem, we have presented a physical-geometrical co-simulation process linking the two models developed under a cooperative simulation interface. We have carried out a 1D-3D co-simulation analysis of a multiphysics mechanism by creating an interface block between the 1D physical model and the 3D geometric model. This instrumentation led us to the solution of optimising the system's essential parameter, namely response time.

This process has given us considerable added value in that it allows us to visualise and analyse the behaviour of the parts in the 3D model, at the same time as simulating the 1D physical model of the system's multiphysical operation.

CHAPTER 6

CONCLUSIONS & OUTLOOK

The main objective of this study is to analyse aeronautical multiphysics systems using a versatile approach to modelling and simulating fighter aircraft braking systems. This approach is intended to contribute both to the study of the behaviour and to the design of this type of system.

Firstly, it was necessary to clarify the concepts of systems engineering and the multiphysics classification of complex systems, such as those involved in mechatronics. Following an analysis of the traditional, sequential and job-based design process, we have supported a contemporary multiphysics design methodology that meets the requirements of concurrent and collaborative engineering. This methodology is ideally suited to multiphysics design, in particular by adopting a V-cycle. This design cycle is essentially based on the development of virtual models and their

validations using simulations at the level of abstraction required for conclusive integration.

In this respect, we have addressed the notion of a model and its levels of abstraction, in order to establish the limits of modelling a multidisciplinary system based on structural decomposition. To this end, an evaluation of the different approaches enabled us to identify the strengths and limitations of each.

The main difficulty in this modelling lies in the inability to link the levels of abstraction of the model of a multiphysical system. This problem of modelling and simulating a multiphysical system is considerably accentuated in the case of aircraft systems, which are becoming increasingly complex.

We have outlined the multi-disciplinary nature of aeronautical systems, such as those powered by hydraulic and electrical generation, as the aircraft's primary sources of power.

In this context, our research focused on the study of fighter aircraft braking systems as critical multiphysics systems, to make our approach viable. We used a process based on systems engineering to analyse the multiphysics operation of these systems through the exploration of system architectures. This investigation included the study of the functional and dimensional parameters of the system components.

One of the first contributions of our research concerns the 1D physical modelling of a multiphysical system. The multivalent approach has enabled us to model multidisciplinary systems. Our proposed 1D physical modelling approach is designed using a truly modular process, adopting component-oriented modelling whose tools are compatible with those of object-oriented programming. Two powerful tools dedicated to this approach for the development of 1D physical models were presented: Simscape / simulink and AMEsim.

This multiphysics approach has been deployed on the two multidisciplinary fighter aircraft braking systems to develop their 1D physical models and those of their sub-systems.

The success of this development is due to the exploitation of the syntheses of the multiphysics study of the operating principle and geometrical parameters, carried out on these two systems instead of writing their calculation algorithms. In fact, we make use of the multiphysics libraries of the basic models of the tools, and if necessary, the development of submodels that can be integrated into our modelling environments. This process can be summarised as follows:

Study of the operating principle and functional parameters of the components that make up the systems by noting the geometric and physical parameters;

Decomposition of the multidisciplinary system into multiphysical subsystems to facilitate modelling of the complete system;

Exploitation of the multiphysics study and the use of modelling environments (AMESim or Simscape) to define the basic sub-models of the libraries of these environments corresponding to each component of the system;

Development of new sub-models for sub-systems that can be integrated into these environments;

Readjusting the parameters of the basic library submodels ;

Joining the sub-models that make up the physical models of components and sub-systems using the acausal task via

ports" which are associated by the exchange of "energy" while respecting the functional specificities of the system;

Interconnection between the subsystem models enables us to develop a 1D physical numerical model of the multidisciplinary aircraft braking system.

This instrumentation has enabled us to develop physical models of multiphysics braking systems and their multidisciplinary components.

In order to validate, analyse and optimise the behaviour of the 1D physical models developed for aircraft braking systems, we simulated the 1D models.

This simulation in the two environments based on the component-oriented approach, offers the possibility of analysing the variation over time of the physical parameters of the sub-systems and the integral aircraft braking system. In fact, with the simulation of the correct operation test required by the aircraft manufacturer, we were able to validate the models developed.

The validation was conclusive following the comparison of the performance requirements recommended by the test with the simulation results of the corresponding parameters, while highlighting the system's operating principle. These 1D physical models can be used to predict the characteristics of the system's behaviour under different operating conditions.

Furthermore, modelling multidisciplinary aeronautical mechanisms has the disadvantage of not being able to visualise the dynamic and kinematic behaviour of the parts in the 3D geometric model at the same time as simulating multiphysical parameters (pressure, voltage, current, flow, etc.).

Our proposal focuses on a coupled dynamic co-simulation process by developing an interface between the 3D geometric model and the 1D physical model.

Co-simulation offers the possibility of visualising and analysing the behaviour of parts in the 3D model under the multi-physics operation of the system while simultaneously compiling the 1D physical model. We have described the process that can achieve an interconnection between the 3D model and the 1D model via an interface block that is used to simulate two types of physical and geometric model simultaneously. In this way we achieve communication between two levels of model abstraction, which is useful for analysing the behaviour of complex mechanisms by considering the parameters of the main phenomena involved in system operation. We implemented this process on a hydromechanical chamber of the aircraft brake distributor. We were able to extract and analyse the changes over time in the dynamic and kinematic parameters of the parts of the geometric model with the presence of hydraulic fluid in the chamber, while simultaneously running the corresponding 1D physical model simulation. This co-simulation enabled us to visualise the reaction of the mechanical parameters of the geometric model parts to the hydromechanical operation of the mechanism.

In the same vein, we have also visualised and analysed the evolution of the force produced by the braking system on the aircraft wheel in the form of an action-reaction between the brake block, which describes the dynamic behaviour of the aircraft wheel on the ground under brake stress.

The co-simulation of the models enabled us to examine the results in order to optimise the behaviour of the aircraft braking system; in particular, to find the optimum solution in terms of response time as an essential parameter of the system's functional efficiency.

To reiterate, our physical modelling approach is applicable to multiphysical systems, particularly aeronautical systems. The viability of hydromechanical and electro-hydromechanical aircraft braking systems has demonstrated this. In addition, the co-simulation process is also valid for systems containing electrical circuits and mechanical mechanisms, but a single hydraulic or pneumatic enclosure. This restriction is due to the difficulty of geometrically modelling the hydraulic fluid or air contained in an interconnection linking two chambers. If we now look to the future, with regard to the prospects linked to the work carried out, several aspects stand out.

The first perspective we have identified stems from a limitation of our model. In fact, it was only possible to experiment and test the model in its entirety in the three experiments contained in the aircraft manufacturer's performance test. It is therefore interesting to look more closely at the following points:

The co-simulation process can be extended in the case of a system of two hydraulic or pneumatic enclosures or chambers. One possible solution is to model the hydraulic pressures at the outlet of the first chamber and the inlet of the second chamber, and their effects, using mechanical mechanisms that can be integrated into the overall geometric model.

Our contribution can be of greater use in the design and dimensioning of braking system components, particularly brake distributors.

Our study approach can be extended to serve as a reference in the development of virtual prototypes for predicting the maintainability of multiphysical aeronautical, automotive or railway components and systems.

The realisation of the different models of the abstraction levels of the systems of an aircraft with the application of the co-simulation process can lead to the development of a numerical model of the aircraft which will be used to simultaneously analyse the behaviour of the various organs and systems and their functional interactions.

The models developed for the aircraft systems were designed using standard environmental conditions, which are prescribed by the aircraft manufacturer's performance test. Given the wide range of conditions in which an aircraft can evolve, it would be interesting to pursue this line of thought by looking at tools and methods for analysing and quantifying the impact of external environmental parameters on the behaviour of aircraft systems.

In the short term, it is already possible to use these different results on numerous industrial cases. It would then be interesting to gauge the reactions of designers, engineers and quality control specialists.

[1] - PENALVA J. M., "La modëlisation par les systemes en situations complexes". *These de*

doctorate, University of Paris XI. Paris 1997

[2] - BREAS M., "Prob^matique de l'intégration de l'Action Individuelle dans les Systemes Technologiques - Application : utilisation de la description des fonctions pour favoriser l'Autonomie". *Doctoral thesis, Institut National Polytechnique de Lorraine.* Nancy 1993

[3] - LE MOIGNE J. L., "La thëorie du Systeme Gënëral: thëorie de la modëlisation". *Collection*

les Classiques du Reseau Intelligence de la Complexite. 1994

[4] - ASHBY ROSS W., "Facets of Systems Science: General System Theory as a new discipline". *Springer Science+Business Media, p 249.* New York 1991

[5] - IEEE, "P1220 Trial-Use Standard for Application and Management of the Systems Engineering Process". *Institute of Electrical and Electronics Engineers Standards Department, inc.* New York 1994

[6] - MEINADIER J. P., "Ingënierie et intégration des systemes". *Hermes Science Publications.*

1998

S. FIORESE, J. P. MEINADIER, '^couvrir et comprendre l'ingënierie systeme'. *Cepadues.* 2012

[8] - F. BENABEN, C. ANTOINE, J. P. PIGNON, J. MAGNIER, M. LARNAC, "Une mëthode d'aide a la functional design of multi-technology technical systems". *Revue Genie Logiciel.* 2001

P. MASSOTTE, "La modëlisation systematique en entreprise". *Hermes, chapter 12.* Paris 1995

[10] - K. PAETZOLD, W. SCHWEIGER, B. SHI, "An approach to the organization of the design phase in the interdisciplinary development process". *Workshop: Intelligent Mechatronic Systems*. Paderborn 2003

[11] - R. ISERMANN, "On the design and control of mechatronic systems-a survey". *IEEE Transactions on Industrial Electronics, vol. 43*. 1996

[12] - W. BOLTON, "Mechatronics". *Harlow, England: Addison Wesley Longman.* 1999

[13] - N. KYURA, H. OHO, "Mechatronics - an industrial perspective". *IEEE - ASME Transactions on Mechatronics, vol. 1*. 1996

[14] - R. COMERFORD, "Mecha...what". *IEEE Spectrum, vol. 31, pp. 46-49.* 1994

[15] - AFNOR, "Technical Report NF E01-010 in Mëcatronique - Vocabulaire". 2008

[16] - J. GAUSEMEIE, R. DUMITRESCU, S. KAHL, D. NORDSIEK, "Integrative development of product and production system for mechatronic products". *Robotics and Computer-Integrated Manufacturing, vol. 27, pp. 772-778.* 2011

[17] - D. DENEUX, "Mëthodes et modeles pour la conception concourante," *Universite de Valenciennes et du Hainaut Cambresis.* 2002

[18] - P. JAGOU, "Concurrent engineering : la maftrise des couts, des delais et de la qualite". *Hermes.* Paris 1993

[19] - VDI, "Verein Deutscher Ingenieure : Entwicklungsmethodik fur mechatronische systeme (design methodology for mechatronic systems)". *Technical Report VDI 2206, Berlin: Beuth Verlag Gmb: Verein Deutscher Ingenieure.* 2004

[20] - J. M. LEGAY, "L'experience et le modele. Un discours sur la methode". *INRA, p. 111.* Paris 1997

[21] - D. VANBERGUE, "Conception de simulation multi-agents: application a la simulation des migrations intra-urbaines de la ville de Bogota". *These de Doctorat, Universite Pierre et Marie Curie - Paris VI.* Paris 2003

[22] - J. DORAN, "Intervening to achieve co-operative ecosystem management: towards an agent based model". *Journal of Artificial Societies and Social Simulation.* 2001

[23] - F. PECHEUX, C. LALLEMENT, A. VACHOUX, "VHDL-AMS and Verilog-AMS as alternative hardware description languages for efficient modeling of multidiscipline systems". *IEEE transactions on Computer-Aided design of integrated Circuits and Systems, vol. 24, pp. 204-225.* 2005

[24] - A. JARDIN, W. MARQUIS-FAVRE, D. THOMASSET, F. GUILLEMARD, F. LORENZ, "Study of a sizing methodology and a modelica code generator for the bond graph tool ms1". *Proceedings of the 6th International Modelica Conference, Bielefeld.* Germany 2008

[25] - R. KAsPER, A. KocH, A. KAsPER, A. WoLF, "Integrierte Entwicklungsumgebung mechatronischer Kfz Komponenten und Kfz-Systeme". *VDI Berichte, pp. 451-465.* 1997

[26] - H. M. PAYNTER, "Bond Graph Methodology: Analysis and design of engineering systems". *The M.I.T. Press, Cambridge, Mass.* 1961

[27] - c. PRINs, "Algorithmes de graphes". *Eyrolles.* 1994

[28] - W. MARQuls-FAVRE, "Contribution a la conception et la modelisation des systemes mecatroniques (et de leurs sous-systemes mecaniques)". *HDR, Institut National des Sciences Appliquees de Lyon, Universite Claude Bernard Lyon I.* 2007

[29] - o. VERLINDEN, "Simulation du comportement dynamique de systemes multi corps flexibles comportant des membrures de forme complexe". *Doctorat en Sciences Appliquees Faculte Polytechnique de Mons.* 1994

[30] - AERIS, "Boeing 737, Ё^иоKe". 2001

[31] - AERIs, "Boeing 737, Auxiliary Power Unit (APU)". 2001

[32] - C. LALAQUE, "Connaissances Aëronefs JAR-FCL Cellule et Systemes". 2002

[33] - AERIS, "Boeing 737, Hydraulics". 2001

[34] - T. MARGER, "Conception d'un distributeur de servocommande hydromëcanique sous criteres de cout et de mixabilrt^". *These n° 432 ParisTech.* 2011

[35] - D. SCHOLTZ, "Aircraft Systems - Reliability, mass Power and Costs". *Presented at the European Workshop on Aircraft Design Education.* 2002

[36] - PUY DE GOYNE T. D., PLAYS Y., LE POURRY P., BESSE J., "Initiation a l^ronautique". *Cepadues.* 2011

[37] - M. SGHAIRI-HAOUATI, "Innovative architectures for flight control systems". *PhD, thesis of the University of Toulouse.* 2010

[38] - DASSAULT AVIATION, "1T-AJ1H-2-12 flight control maintenance manual". 1988

[39] - NORTHROP, "TO 1F-5E-4-2 Technical Manual Flight Control Systems". 2006

[40] - DASSAULT AVIATION, "1T-AJ1H-2-13 maintenance manual aërofreins vol". 1988

[41] - J. VEAUX, "Les trains d'atterrissage et les systemes asso^s Centre des hautes ëtudes de Division Histoire de l'armement". 2006

[42] - DASSAULT AVIATION, "1T-AJ1H-2-13 flight landing gear maintenance manual". 1988

[43] - DASSAULT AVIATION, "1T-AJ1H-2-14 manuel d'entretien installation hydraulique vol". 1988

[44] - A. TURIOT, M. GENOUILLE-DALFORT, "L'informatique : une aide a la conception des atterrisseurs a tous les niveaux vol". *L'Aeronautique et l'Astronautique, No. 56*. July 1975

[45] - NORTHROP, "To 1F-5E-2-20-7 Technical Manual Landing Gear Systems Aircraft". 2006

[46] - NORTHROP, "F-5 Technical Digest journal". 1975

[47] - F. CHEVILLOT, "Nonlinear transient vibrations and coexistences of multi-instabilities induced by friction in an aircraft braking system". *D. thesis, Ecole Centrale de Lyon*. 2009

[48] - DASSAULT AVIATION, "1T-AJ1H-2-13 manual d'entretien systeme freinage vol". 1988

[49] - S. STEINKELLNER, "Aircraft Vehicle Systems Modeling And Simulation Under Uncertainty". *Thesis Linkoping Studies In Science And Technology*. 2011

[50] - NORTHROP, "TO 1F-5E-2-1-1, Airplane General Technical Manual". 2005

[51] - NORTHROP, "TO 1F-5F-2-20-3 Pneudraulics Troubleshooting". 1996

[52] - NORTHROP, "TO 1F-5E-2-20-7, Landing Gear Systems Technical Manual". 2006

[53] - NORTHROP, "TO 1F-5E-2-20-3, Pneudraulics, chap. 1F-5AB-2-8-29D, Wheel Brake Control Valve Technical Manual". 2000

[54] - NORTHROP, "TO 1F-5E-2-20-3, Pneudraulics, chap. 1F-5AB-2-8-4F, Wheel Brake Control Valve Technical Manual". 2000

[55] - NORTHROP, "TO 1F-5A-2-8, Wheel Brake System System Analysis". 2006

[56] - NORTHROP, "TO 1F-5E-2-20-3, Pneudraulics, chap. 1F-5AB-2-8-5E, Wheel Brake Control Valve Operational Modes". 2000

[57] - DASSAULT AVIATION, "Notice Technique Distributeur Double vol". 1979

[58] - DASSAULT AVIATION, "Notice Technique Distributeur Secours vol". 1984

[59] - DASSAULT AVIATION, "Notice Technique Ëlectrovalve Freinage vol". 1989

[60] - DASSAULT AVIATION, "Notice Technique gënëratrice tachymëtrique vol". 1979

[61] - DASSAULT AVIATION, "Notice Technique boitier de regulation Modistop vol". 1990

[62] - DASSAULT AVIATION, "1T-AJ1H-2-15 Maintenance manual installation ëlectrical flight". 1980

[63] - DASSAULT AVIATION, "Notice Technique Bloc frein vol". 1986

[64] - A. NAAMANE, M. RADOUANI, A. SAKA, "Modeling of aircraft braking system, component

oriented approach". *IJRRMDS, International Journal of Research and Reviews in Mechatronic Design and Simulation, vol. 1, pp. 43-48*. September 2011

[65] - A. NAAMANE, M. RADOUANI, B. EL FAHIME, A. SAKA, "Modëlisation multi-physique d'un aircraft brake distributor". vn .louriiees d'Etudes Techniques. *The International Congress for Applied Mechanics, JET'2012, p. 8*. Marrakech 2012

[66] - A. NAAMANE, M. RADOUANI, A. SAKA, "Modeling of aircraft braking system, component oriented approach". *CPI'2011, 7th International Conference on Integrated Design and Production, p. 14*. Oujda 2011

[67] - NORTHROP, "TO 1F-5E-2-7-5, Test wheel brake system. Changel-Emergency Mode, Change2- Normal Mode". 1987

[68] - A. NAAMANE, M. RADOUANI, S. AKROUT, B. EL FAHIME, J. Y. CHOLEY, "Modeling of mechatronic aircraft braking system". *IJMRAE, International Journal of Multidisciplinary Research and Advances in Engineering. Ascent International Journals, vol. 5, pp. 37-48*. April 2013

[69] - B. EL FAHIME, "Contribution au Toterancement des Systemes Mëcatroniques". *These de doctorat, Universite Sidi Mohamed Ben Abdellah - Faculte des Sciences et Techniques, CED "Sciences et Techniques de Hngenieur"*. Fes 2012

[70] - N. MHAITI, A. NAAMANE, M. RADOUANI, B. EL FAHIME, Y. AOURA, "Multidisciplinary Modeling and Co-simulation of an Aircraft Brake Control Valve". *CPI'2013, 8th International Conference on Integrated Design and Production*. Tlemcen - A1дёпе 2013

[71] - A. NAAMANE, N. MHAITI, R. RADOUANI, B. EL FAHIME, "Co-operative geometricphysical simulation of a multidisciplinary aircraft mechanism," *IJRAME, International Journal of Research in Aeronautical and Mechanical Engineering, vol. 1, pp. 274-280*. November 2013

[72] - LMS Imagine.Lab AMESim Rev 13, Siemens. *www.lmsintl.com*

Appendices

Appendix 1

Distributor spring stiffness :

The spring stiffnesses of the spool valve have been determined by means of a test. Stiffness, by the application of a force (Table 5).

Tulip valve	0.251 Nm^{1}	Pressure sensing valve	4.61 8 Nm^{-1}
Stem	0.192 Nm^{1}	Measuring tap	0.114 Nm^{1}
Master cylinder piston	0.063 Nm^{-1}	Large piston chamber	29,421 Nm^{1}
Shuttle valve	0.624 Nm^{-1}	Small piston chamber	25,059 Nm^{1}

Table 5. *Spring stiffness of the spool valve*

Dispenser materials

The distributor is made mainly of aluminium-copper alloy AlCu4MnSi (AW-2017A), which has good mechanical properties, particularly in the tempered or matured state. Its mechanical properties (density 2.96g/cm^3 ; modulus of elasticity 74000MPa; Poisson's ratio 0.33; yield strength 260MPa; fracture toughness 390MPa) make it a popular choice for aerospace applications. It also has good heat resistance and peeling properties. On the other hand, it has poor resistance to corrosion in corrosive atmospheres due to the presence of copper. In general, it is used for parts subject to stress. In addition, 2017, formerly known in France as duralumin, is used for its good machining properties.

Appendix 2

CREATION OF A 1D MODEL OF AN AERONAUTICAL COMPONENT

CAN BE INTEGRATED INTO THE AMESIM INTERFACE

If there are no models of aeronautical components in the libraries, a basic model can be created using the AMESim tool by following the steps described below.

Creation of our component icon on AMESet

In AMESet, the process is launched from the **Icons** menu. The steps to follow are :

Step 1 Determine whether you need to create a new

category. If you already have a suitable category, go on to step 3.

Step 2 Create an icon for the new category.

Select Modeling > Category settings > Add category

Select where you want to put your category:

Figure 98. *Location of our category*

If this location is not in AMESet this message will appear:

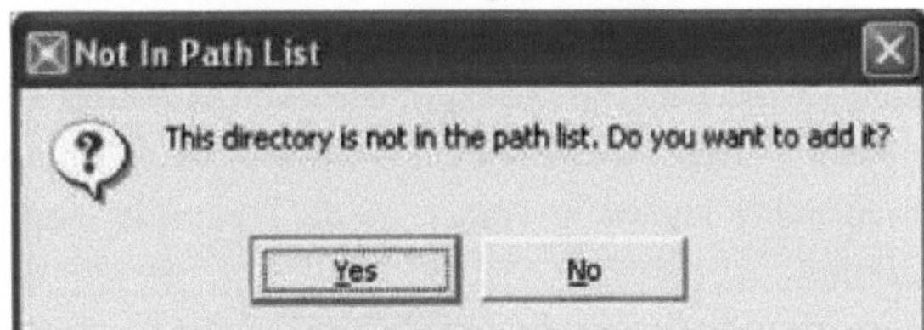

Figure 99. *Message requesting validation of the location*

Step 3 Create a SOUPAPE_TULIPE icon

Draw the SOUPAPE_TULIPE icon, and don't forget the port.

Add this icon directly to the tutorial/icons location.

The name of the icon is: sOuPAPE_TuLiPE.xbm

Step 4 Check that our category exists, and that it contains the drawn icon.

Step 5 Choose a colour for our category: Options > Color preferences

Creation of the sub-model of our SOUPAPE_TULIPE

How will the sub-model work? We've already decided that we're going to have 4 external variables associated with the port. We're going to work with international units and give the variables common names. In conclusion, we have the following variables:

Explicit state: 'the main pressure of the sOuPAPE_TuLiPE' [bar].

Explicit state: 'the secondary pressure of the sOuPAPE_TuLiPE' [bar].

Basic variable (input) : 'the main force on the sOuPAPE_TuLiPE' [N]

Basic variable (input): 'the secondary force on the sOuPAPE_TuLiPE' [N].

Mass of TULIPE_SPRING ' [kg]

Geometric dimensions ' [m]

Step 1 Start AMESet, and choose the icon of the SOUPAPE_TULIPE from its category and select a new submodel.

Step 2 Give the port 4 external variables, and define them:

AMESim imposes the following rule: outputs before inputs before inputs by default.

To correct the order of the external variables, right-click on the variable you want and choose move up or move down.

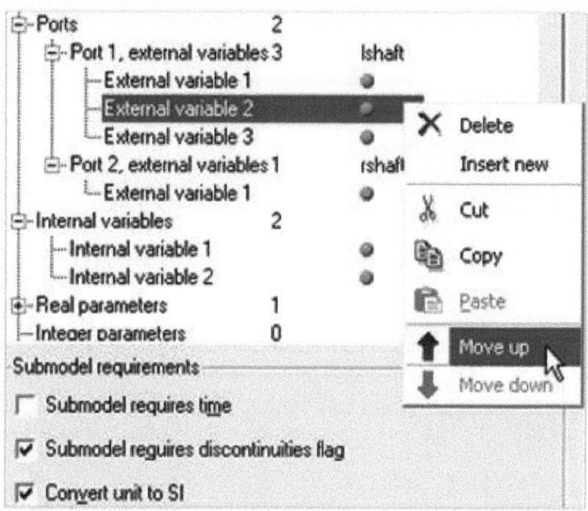

Figure 100: *Changing the order of external variables*

Step 3 Choose a number of real parameters of 1 (mass) for example, and define it as follows:

Field	Value
Title	Mass of ...
Unit	kg
Variable name	mass
Min.	0.001
Default	1
Max	10

Table 6. *Real parameter (mass)*

Step 4 Give a description of the sub-model of the SOUPAPE_TULIPE :

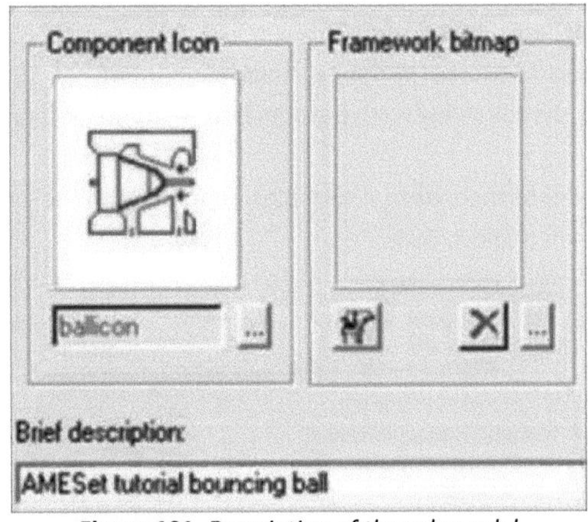

Figure 101. *Description of the sub-model*

Step 5 Save the sub-model with the name **SOUPAPE_TULIPE**

Step 6 Write the code to complete the sub-model

```
C >>>>>>>>>>>>Calculation Function Executable Statements.
/* Submodel SOUPAPE_TULIPE skeleton created by AME Submodel editing
utility
   jeu. 13. déc. 10:56:52 2012 */
#include <math.h>
#include <stdio.h>
#include <stdlib.h>
#include "ameutils.h"
/* ****************************************************************
TITLE :
 ------------------------------------------------------------------
--------------
DESCRIPTION :
 ------------------------------------------------------------------
--------------
USAGE :
 ------------------------------------------------------------------
--------------
PARAMETER SETTINGS :
 ------------------------------------------------------------------
--------------
REVISIONS :

******************************************************************
************** */
/* >>>>>>>>>>>>Insert Private Code Here. */
```

```
/* <<<<<<<<<<<<End of Private Code. */
void soupape_tulipein_(int *n)
{
   int loop, error;
/* >>>>>>>>>>>>Extra Initialization Function Declarations Here. */
/* <<<<<<<<<<<<End of Extra Initialization declarations. */
   loop = 0;
   error = 0;
/* >>>>>>>>>>>>Initialization Function Check Statements. */
/* <<<<<<<<<<<<End of Initialization Check Statements. */
   if(error == 1)
   {
      amefprintf(stderr,  "\nWarning  in  SOUPAPE_TULIPE  instance
%d.\n", *n);
   }
   else if(error == 2)
   {
      amefprintf(stderr, "\nFatal error in SOUPAPE_TULIPE instance
%d.\n", *n);
      amefprintf(stderr, "Terminating the program.\n");
      AmeExit(1);
   }
/* >>>>>>>>>>>>Initialization Function Executable Statements. */
/* <<<<<<<<<<<<End of Initialization Executable Statements. */
}
/*  There are 4 ports.
   Port 1 has 0 variables:
   Port 2 has 0 variables:
   Port 3 has 0 variables:
   Port 4 has 0 variables:
*/
/*  There are 0 internal variables.
*/
void soupape_tulipe_(int *n)
{
   int loop;
/* >>>>>>>>>>>>Extra Calculation Function Declarations Here. */
/* <<<<<<<<<<<<End of Extra Calculation declarations. */
   loop = 0;
/* >>>>>>>>>>>>Calculation Function Executable Statements. */
/* <<<<<<<<<<<<End of Calculation Executable Statements. */
}
```

Step 7 Compile the code and correct any errors.

Step 8 Check that you can use the sub-model in AMESim

Check that the sub-model can be connected to the brown icons in the Hydraulic Compenent Design library. The sub-model ports must be able to connect to others that provide pressure and flow.

force. Try these connections:

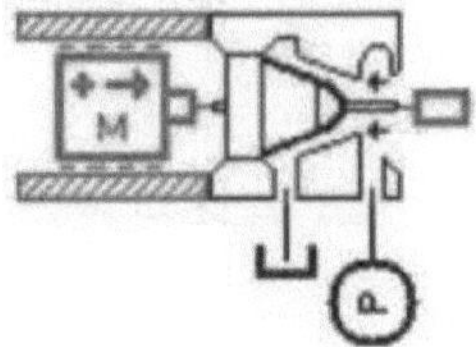

Figure 102. *Checking the SOUPAPE_TULIPE connection*

Connect the TUBE VALVE to a zero force source.

Run the simulation with the default parameters, and you can see how the TULIP_VALVE behaves.

Appendix 3

AMESIM BRAKE PEDAL MODEL 72]

```
******************************************************************
TITLE : BRAKE PEDAL
 ------------------------------------------------------------------
--------------
DESCRIPTION :
    This is a simple model of a brake pedal ; it computes the force
that is applied on the rod from the driver input on the pedal which
can be either a  force or a displacement.
    The relation between output and input can be defined by two
different ways :
      - as a datafile
      - as an expression
 ------------------------------------------------------------------
--------------
USAGE :
    Use this element to model a brake pedal ; it must be located
just before the model of power brake booster.
 ------------------------------------------------------------------
--------------
PARAMETER SETTINGS & EQUATIONS :
    In both cases - table or an expression setting - the output
force on the rod is asked in function of the input force on the
pedal.
 ------------------------------------------------------------------
```

```
/* >>>>>>>>>>>>Insert Private Code Here. */
/* <<<<<<<<<<<<End of Private Code. */

/* There are 2 real parameters:
   xminslope slope for x < xmin [null]
   xmaxslope slope for x > xmax [null]
*/
/* There are 5 integer parameters:
   spline interpolation type
   lmode  linear data out of range mode
   bcond  boundary conditions
   cmode  cubic data out of range mode
   disc   discontinuity handling
*/
/* There is 1 text parameter:
   filexp filename or expression for output force Fout = f(Fin)
Fout in N , Fin (N or m or %)
*/
void poigne01in_(int *n, double rp[2], int ip[5], char *tp[1]
      , int ic[1])
{
   int loop, error;
/* >>>>>>>>>>>>Extra Initialization Function Declarations Here.
*/
   int dim = 1,           /* table dimension = 1 */
       num_tokens = 1,    /* number of variables in the expression
= 1 */
       int_mode,          /* interpolation mode */
       ext_mode[2],       /* extrapolation mode */
       bcond_array[2];
   double slopes[2];
   static char *var[] = { "Fin" }; /* name of the variable */
/* <<<<<<<<<<<<End of Extra Initialization declarations. */
   int spline, lmode, bcond, cmode, disc;
   double xminslope, xmaxslope;
   char *filexp;
   spline     = ip[0];
   lmode      = ip[1];
   bcond      = ip[2];
   cmode      = ip[3];
   disc       = ip[4];
   xminslope  = rp[0];
   xmaxslope  = rp[1];
   filexp     = tp[0];
   loop = 0;
   error = 0;
/*
   If necessary, check values of the following:
   rp[0..1]
*/
/* >>>>>>>>>>>>Initialization Function Check Statements. */
   if (spline == 1)
   {
```

```
      bcond_array[0] = lmode;
      bcond_array[1] = lmode;
   }
   else if (spline == 2)
   {
      bcond_array[0] = bcond;
      bcond_array[1] = bcond;
      slopes[0]      = xminslope;
      slopes[1]      = xmaxslope;
      ext_mode[0]    = cmode;
      ext_mode[1]    = cmode;
   }
/* int_mode = 1 : constant interpolation
   int_mode = 2 : linear interpolation
   int_mode = 3 : cubic interpolation */
   int_mode = spline + 1;
   ic[0] = strinit_(filexp, &dim, var, &num_tokens, &int_mode,
bcond_array, ext_mode, slopes, &disc);
   if (ic[0] < 0)
   {
      /* An error occured during spline initialization. */
      amefprintf(stderr, "\nCannot initialize spline %s.\n",
filexp);
      error = 2;
   }
/* <<<<<<<<<<<<End of Initialization Check Statements. */

/*  Integer parameter checking:   */

   if (spline < 1 || spline > 2)
   {
      amefprintf(stderr, "\ninterpolation type must be in range
[1..2].\n");
      error = 2;
   }
   if (lmode < 1 || lmode > 2)
   {
      amefprintf(stderr, "\nlinear data out of range mode must be
in range [1..2].\n");
      error = 2;
   }
   if (bcond < 1 || bcond > 3)
   {
      amefprintf(stderr, "\nboundary conditions must be in range
[1..3].\n");
      error = 2;
   }
   if (cmode < 1 || cmode > 2)
   {
      amefprintf(stderr, "\ncubic data out of range mode must be
in range [1..2].\n");
      error = 2;
   }
```

```
   if (disc < 1 || disc > 2)
   {
      amefprintf(stderr, "\ndiscontinuity handling must be in
range [1..2].\n");
      error = 2;
   }
   if(error == 1)
   {
      amefprintf(stderr, "\nWarning in POIGNE01 instance %d.\n",
*n);
   }
   else if(error == 2)
   {
      amefprintf(stderr, "\nFatal error in POIGNE01 instance
%d.\n", *n);
      amefprintf(stderr, "Terminating the program.\n");
      AmeExit(1);
   }
/* >>>>>>>>>>>>Initialization Function Executable Statements. */
/* <<<<<<<<<<<<End of Initialization Executable Statements. */
}
/*  There are 2 ports.
   Port 1 has 1 variable:
      1 f1     pedal output torque at port 1 [Nm] multi line
macro 'poigne01_macro0_'
   Port 2 has 1 variable:
      1 f2     pedal input at port 2 [null] basic variable input
*/

/*  There are 0 internal variables.
*/
#if 0
/* THE CALCULATION FUNCTION WILL NOT BE CALLED. */
void poigne01_(int *n, double *f1, double *f2, double rp[2], int
ip[5]
      , char *tp[1], int ic[1])
{
   int loop;
/* >>>>>>>>>>>>Extra Calculation Function Declarations Here. */
/* <<<<<<<<<<<<End of Extra Calculation declarations. */
   int spline, lmode, bcond, cmode, disc;
   double xminslope, xmaxslope;
   char *filexp;
   spline     = ip[0];
   lmode      = ip[1];
   bcond      = ip[2];
   cmode      = ip[3];
   disc       = ip[4];
   xminslope  = rp[0];
   xmaxslope  = rp[1];
   filexp     = tp[0];
   loop = 0;
/*
```

```
   Set all submodel outputs below:
*/
/* >>>>>>>>>>>>Calculation Function Executable Statements. */
/* <<<<<<<<<<<<End of Calculation Executable Statements. */
}
#endif
extern double poigne01_macro0_(int *n, double *f2, double rp[2]
      , int ip[5], char *tp[1], int ic[1])
{
   double f1;
   int loop;
/* >>>>>>>>>>>>Extra Macro Function macro0 Declarations Here. */
   int success;
/* <<<<<<<<<<<<End of Extra Macro macro0 declarations. */
   int spline, lmode, bcond, cmode, disc;
   double xminslope, xmaxslope;
   char *filexp;
   spline     = ip[0];
   lmode      = ip[1];
   bcond      = ip[2];
   cmode      = ip[3];
   disc       = ip[4];
   xminslope  = rp[0];
   xmaxslope  = rp[1];
   filexp     = tp[0];
   loop = 0;
/*
   Define and return the following macro variable:
   f1         = ??;
*/
/* >>>>>>>>>>>>Macro Function macro0 Executable Statements. */
   success = streval_(&ic[0], f2, &f1);
   if (success == 0)
   {
      amefprintf(stderr, "\nCannot evaluate expression or read
filename %s.\n", filexp);
      amefprintf(stderr, "\nFatal error in BRAKE PEDAL instance
%d.\n", *n);
      amefprintf(stderr, "Terminating the program.\n");
      AmeExit(1);
   }
/* <<<<<<<<<<<<End of Macro macro0 Executable Statements. */
   return f1;
}
```

MATHEMATICAL REPRESENTATION OF A COMPONENT [72]

1- **Hydraulic port**

Output flow calculation equations

- Dh = 4 . A / P
- Λ = Dh . v / μ = $\sqrt{2\Delta p/\rho}$. Dh / μ
- Cq = Cqm . Tan (2 . λ / λcrit)
- Q = Cq . A $\sqrt{2\Delta p/\rho}$

Dh : Diamètre hydraulique
A : Surface en coupe transversale de l'orifice
P : Périmètre mouillé de section transversale
Cq : Coefficients de débit
λ : Nombre de Reynolds
λcrit: Nombre de Reynolds du transfert de laminaire-turbulent
μ : Viscosité
ρ : Densité
Q : Débit

- **Code C de génération des différentes fonctions associées**

```
extern double orif4f_
 (
 #ifndef _NO_PROTO
 double *p1, double *p2, double *area, double *hd,
 int *indexlambda, double *cq, double *lambda, double *flv
 #endif
 );
 orif4f_(p1, p2, area, hd, indexlambda, cq, lambda, flv);
```

- **Ajustement des paramètres d'entrée de l'orifice**

Title	Value	Unit
index of hydraulic fluid	0	
parameter set for pressure drop	orifice diameter/maximum flow coefficient pair	
equivalent orifice diameter	1.5	mm
maximum flow coefficient	0.7	null
critical flow number (laminar -> turbulent)	1000	null

Adjustment of orifice input parameters

2- **Double chamber spring hydraulic cylinder**

Description

The model includes pressure dynamics in both chamber volumes, viscous friction and internal piston leakage. An external velocity and displacement (Inputs) must be supplied to the Port 3 shaft by another model and the force generated is the Output parameter. The two flow ports require flow as Input and pressure as Output. Due to an enumeration parameter the cylinder can be used in two different modes with or without the "use of an initial displacement":

With : the initial movement of the external component connected to port 3 is ignored, and initialisation is performed internally in the piston;

Without: the initial position of the actuator is determined by the external component connected to port3.

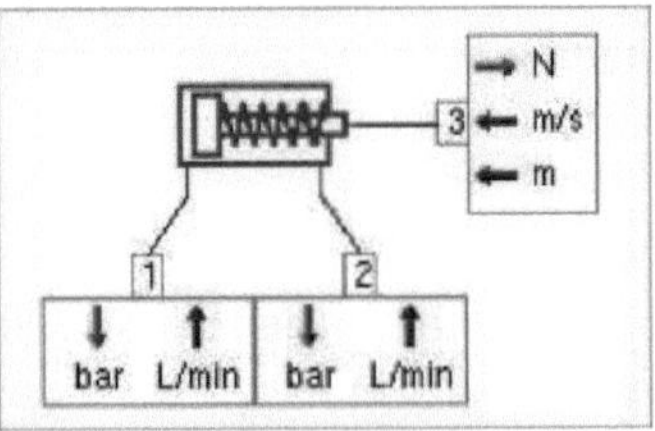

External variables

	Title	I Variable	u I Unite I	Type	I In/Out I	Default value
Port 1						
1	pressure port 1	p1	bar	Explicit state	output	0
2	port 1 throughput	q1	L/min	base variable	input	-
Port 2						
1	pressure port 2	p2	bar	Explicit state	output	0
2	port 2 throughput	q2	L/min	base variable	input	-
Port 3						
1	force exerted by the cylinder rod	frod	N	base variable	output	-
2	Input speed	v	m/s	base variable	input	-
3	MOVING IN	x	m	base variable	input	-

Internal variables

Title	Variable	Unit	Type	by default
Deplacement of piston	xact	m	base variable	-
Leakage rate	qleak	L/min	base variable	-
Hydraulic dissipation activity (leakage)	actRleak	J	Explicit state	0
Hydraulic dissipation power (leakage)	powerRleak	W	variable power	-
hydraulic dissipation energy (leakage)	energyRleak	J	Explicit state	0
Hydraulic capacity activity (chambrel)	actCchamberl	J	Explicit state	0
Hydraulic capacity (chamberl)	powerCchamberl	W	power variable	-
Energy from hydraulic capacity (chambrel)	energyCchamberl	J	Explicit state	0
Hydraulic capacity activity (chamber2)	actCchamber2	J	Explicit state	0
Hydraulic capacity (chamber2)	powerCchamber2	W	Power variable	-
Hydraulic capacity energy (chamber2)	energyCchamber2	J	Explicit state	0
Mechanical dissipation activity (friction)	actRfrict	J	Explicit state	0
Mechanical dissipation power (friction)	powerRfrict	W	Power variable	-
mechanical dissipation energy (friction)	energyRfrict	J	Explicit state	0
Activation of mechanical capacity (spring)	actCspring	J	Explicit state	0
Power of mechanical capacity (spring)	powerCspring	W	Power variable	-
Energy from mechanical capacity (spring)	energyCspring	J	Explicit state	0
Water source activity (patm)	actSpatm	J	Explicit state	0
Hydraulic source power (patm)	powerSpatm	W	Power variable	-
Energy from hydraulic sources (patm)	energySpatm	J	Explicit state	0

Actual parameters

Title	Variable	Unit	Default value
Piston displacement	xact0	m	0
Piston diameter	diamp	mm	25

Stem diameter	diamr	mm	12
Length of stroke	stroke	m	0.3
Volume break-even point port 1	deadl	cm**3	50
Break-even volume port 2	dead2	cm**3	50
Viscous friction coefficient	Cvisc	N/(m/s)	0
Leakage coefficient	Cleak	L/min/bar	0
Spring load	preld	N	0
Spring return constant	Ksprng	N/m	10000
Spring return constant	k	N/mm	100000
End-of-travel damping coefficient	Cdamp	N/(m/s)	100000
End-of-stroke deformation with damping rate is fully effective	distdef	mm	0.001

Integer parameters

1 Title	Variable 1	Default value1
Hydraulic fluid index	indexf	0
Use of initial move	usedispl	1: yes

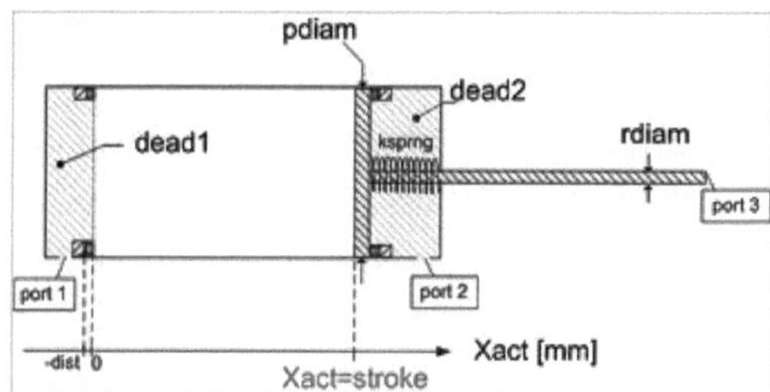

Sketches and equations

> **Mechanics**

The force exerted by the rod due to pressure, viscous friction and spring force is :

$$f_{rod} = p_1 A_1 - p_2 A_2 + v.c_{visc} - P_{reld} - k_{spring}.X_{act}$$

With : $A_1 = \pi.diam_p^2 / 4$ et $A2 = \pi.(diam_p^2 - diam_r^2) / 4$

Where: A1 is the cross-sectional area of the piston on which activates pressure P1 ;

A2: the annular surface on which pressure P2 is applied.

An additional force will be applied if the piston comes into contact with the end stop. Consequently, there are three distinct modes of operation: ■ On lower end fitting if $x_{act} < 0$: the following expression is added to f_{rod}

$$-k.x_{act} + v.cd_{touse}$$

Où : $cd_{touse} = k.[1 - \exp(x_{act} / distdef)]$

Or :

■ on the upper bend if $x_{act} > x\ k_{stroe}$: Similarly, the following term is added to f_{rod}

$$-k.(x_{act} - stroke) + v.cd_{touse}$$

Où : $cd_{touse} = k.[1 - \exp((x_{act} - stroke) / distdef)]$

Or :

■ free end fitting : frod is unmodified

si $x>0$ $f_{rod}=k.x+r_{eff}.v$; avec : $r_{eff}=r.(1-\exp(-x/e))$

si $x\leq 0$ $f_{rod}=0$

if

if

```
k: contact stiffness
x : relative displacement between the two moving bodies reff
   effective contact damping coefficient
v: relative speed between the two moving bodies e: deformation
```

Thus, the value of the force applied to the load depends on the *Xact* of piston displacement in the following two cases:

■ with "use of initial displacement" :

The X_{acto} parameter is used to define the initial piston displacement.

On the first call to the calculation function, the offset between this displacement and the initial displacement of the external component at port 3 is calculated and stored in the intermediate variable C :

$$C=x_{act0}+x_{firstcall}$$

The value of C is retained, and the value of X_{act} can then be determined from x during the simulation using :

$$x_{act}=C-x$$

Consequently, the absolute positions of the piston (x_{act}) and the component external to port 3 (x) are independent. Only their relative displacements are equal.

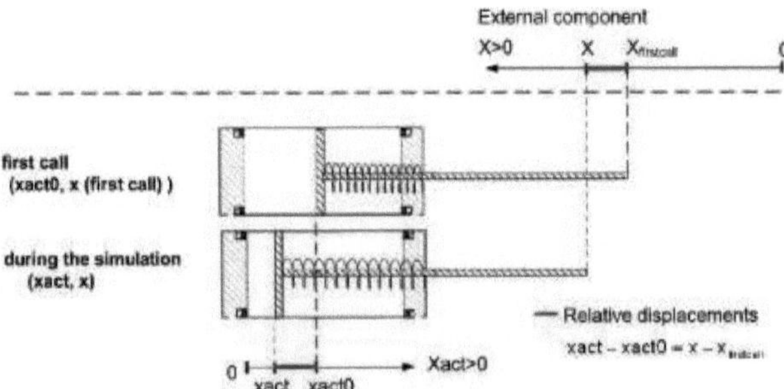

■ Without "use of initial displacement":

In this case, the initial displacement of the external component at port 3 (x) is the same as the piston displacement x_{act}.

$x_{act}=-x$

■ **Hydraulics**

Leakage rate $q_{leak} = (p_1 - p_2).leak$

- Calculating the pressure derivative at port 1

The volume of fluid in the chamber corresponding to port 1 is :

$$vol_1 = A_1.x_{act} + dead_1$$

The flow in the chamber at port 1 is :

$$inflow_1 = q_1 - q_{leak} - v.A_1.\rho(p_1)/\rho(0)$$

Calculating the pressure derivative at port 2

The volume of fluid in the chamber corresponding to port 1 is :

$$vol_1 = A_1.\ (stroke - x_{act}) + dead_1$$

The flow in the chamber at port 2 is:

$$inflow_2 = q_2 - q_{leak} - v.A_2.\rho(p_2)/\rho(0)$$

Printed by Books on Demand GmbH, Norderstedt / Germany